PROCESS PUMP SELECTION
– A SYSTEMS APPROACH

Process pump selection
– a systems approach

Edited by
JOHN DAVIDSON, MBA, BSc (Hons), CEng, FIMechE, FIPlantE

The Publishers are not responsible for any statement made in this publication. Data, discussion and conclusions developed by authors are for information only and are not intended for use without independent substantiating investigation on the part of potential users.

Typeset by Santype International Ltd., Salisbury, Wilts.
Printed in England at The Alden Press, Oxford

CONTENTS

v

Appendices

PREFACE

An engineer employed in the process industries and faced with responsibilities for the specification, selection, and purchase of process equipment has a daunting task. If for any reason the equipment he selects fails to meet the requirements of the plant or process the maximum claim in law against the equipment supplier is for the price paid for the equipment. In contrast the real cost of plant downtime and lost production is likely to exceed the equipment price manyfold. Thus the normal rule for the relationship between purchaser and seller of *caveat emptor* (may the buyer beware) applies in this particularly critical sphere of activity.

With some plant and equipment there are a series of national specifications which lay down design and constructional requirements, e.g., BS 5500 for the design and construction of pressure vessels, and others for piping materials, flanges, etc. In these cases the engineer has some well defined bench marks against which to check his own experience and judgement when purchasing equipment. For process machinery in general, and for pumps in particular, the benefits of national specifications covering constructional requirements hardly exist. The petrochemical industry bases its purchase of centrifugal pumps on the American Petroleum Institute Standard API 610, but for the remainder of the process-industry companies there is no similar standard. The pump purchasers situation is exacerbated by the fact that there are no published design guides to help chart the way through the complex process of specifying the pump requirements.

Most engineers given this responsibility learn from their own experience, however limited, as they progress from one job (or task) to another. In larger organizations such as the oil companies, larger chemical companies, and public sector bodies such as steel, gas, and electricity, individual engineers or groups of engineers specialize and, over a period of time, codify their collective experience in the form of in-house specifications or design guides. Others, less fortunate perhaps, place themselves almost entirely in the hands of suppliers or plant contractors, *caveat emptor* notwithstanding.

The strength of this publication is that it represents the distilled experience of a team of engineers working for one large process company over a period of more than 20 years. It is a classic example of experiential learning, where experience with hundreds of pumping installations has been categorized, analysed, and the essential lessons extracted. These essential lessons have then been synthesized to provide a series of empirical formulae, design guides, and codes of practice for pump users. It is the very essence of professional engineering practice.

The publication is aimed at engineers and others working for user organizations in the process and service sector industries, but will almost certainly have wider appeal. It should not only be of great assistance to engineers faced with the responsibility for the specification, selection, and procurement of pumps, but should also provide pump manufacturers with a great insight into the problems facing pump users and plant designers. It explains how uncertainties which arise during the plant design process are dealt with when decisions need to be taken on pump selection. It should also prove to be invaluable as a basis for the training of young engineers. Against this background it is realized that pump manufacturers may find some of the empirical guidelines laid down here more restrictive than particular ranges of pumps would apparently satisfy. This may be particularly true in certain highly competitive market areas. However, the importance of selection based on the overall suitability for service, including the desired level of reliability, cannot be overemphasized. One can only reiterate that these guidelines are based on the operational experience of large numbers of pumping installations over many years on a diverse range of duties and process plants. This experience takes account of the essential need for plant reliability and all the exigencies of process plant operations.

It is not the purpose of these guidelines to inhibit the development of pump designs beyond the 'state of the art'. Such improvements may be derived as a result of new design developments or as an extension to an existing range in the context of the past experience of a user. The answer to the apparent contradiction between development and reliability is a fairly simple one. Reliability is such an important factor in all plant and equipment for process plants that any new pump design must be assessed on the basis of acknowledged reliability analysis and criteria such as the Duane method for learning curves. Thus in order to give statistically acceptable results which would meet the reliability requirement implied in API 610 of 20 000 hours continuous operation, one would need to see four pump installations on similar duties for four years. There is clearly a place for trial installations in process plants and for cooperation between supplier and major user.

A final, but essential, point for readers and potential users of these guidelines is the question of updating the many empirical constants used in formulae and tables as more information becomes available. As the approach is largely based on parametric analysis of data collected by one large process company, with a 'rolling validation' taking place as more information has become available, it would be surprising if the experience of other large organizations did not differ to some degree or other. In a publication such as this, how does one cope with the problem of updating? A number of possible courses of action spring to mind, viz.

(a) For the more sophisticated large user, either the information and methods contained herein can be used to modify their own approach, or data from their own experience can be used to update or modify the guidelines to suit their particular range of applications.

(b) Where (a) is not possible, and for smaller user organizations, information/comment may be fed back to the Editor for possible incorporation into subsequent editions.

(c) The publication may provide a basis for an independent national design guide/specification which would have all the mechanisms for updating built into it.

ACKNOWLEDGEMENTS

The editor wishes to thank the Directors of Imperial Chemical Industries PLC for permission to publish material which forms the bulk of this publication. It is appropriate also to express thanks to the engineers who have contributed to the preparation of this base material and in particular to thank Mr Alan Axford who has been an important figure in this field over a period of more than 20 years. Mr Axford has also played a key part in the preparation of this publication and has given his time and expertise unsparingly.

Again grateful thanks are due to Mr Trevor Cuerel of British Petroleum Engineering Department for both his editorial and detailed technical comments.

In addition the editor wishes to thank the Engineering Equipment and Materials Users Association for permission to reproduce Section Three of the EEUA Handbook, *A Guide to the Selection of Rotodynamic Pumps* as the Addendum to this publication.

'Dimidium facti qui coepit habet: sapere aude'

To have begun is half the job: be bold and be sensible

HORACE

PART ONE

The decision-making process for the specification, selection, and integration of pumps into process systems

CHAPTER 1. BASIC APPROACH TO PUMP SPECIFICATION AND SELECTION

Success in the application of engineering concepts in the process industries is invariably based on a systems approach to the problems presented. One must look beyond the immediate manifestation of the problem and consider the system as a whole. In considering process equipment it is necessary to consider how the whole system is affected by the installation, and to check how the system interacts with the piece of equipment. In other words, one must see how the equipment can be integrated into the process in order to provide a safe, efficient, and reliable system.

Pumps provide the 'heart' of most process systems, and are the means by which the 'life-blood' of the process is transferred from one part of the system to another. What are the factors that need to be considered in order to effectively integrate pumps into process systems? Clearly the basic factors of delivery head and quantity of fluid to be transferred have a significant effect on the choice of the pump to be used. The properties of the fluid to be transferred can also have a significant effect on the choice of pump, varying from viscosity effects to the effect of differing sizes and quantities of solid particles carried by the fluid. Similarly, the size of storage vessels and their fluid residence times, and the type of flow control system adopted, can have an effect on the choice of pumping installation. The choice of pump itself then interacts with the system. If a positive displacement pump is selected the effects of pressure pulsations need to be accounted for, as do the requirements for pressure relief. For centrifugal pump installations the requirements for adequate inlet arrangements are critical in order to avoid cavitation; and the shape of the pump characteristic affects both the ability of the system to operate with pumps in parallel and the pressure rating of the system.

It is therefore essential to realise that the whole

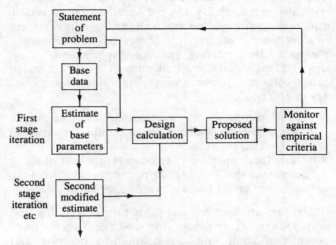

Fig. 1.1. The iterative process of plant and system design

process of plant and system design is one based on iteration. One must first start by making an estimate based on process and/or other preliminary data. This estimate may be a numerical value for a parameter, or a set of parameters, or it may be in the form of a line diagram or layout, depending on the nature of the problem. A solution to the original problem is then sought using the estimate as a starting point for the application of design calculations and/or other design techniques. The solution is then checked for acceptability based on criteria culled from experience. If the solution is not acceptable then a second stage of iteration is carried out based on a second modified estimate of the base parameter(s), and so on, until a satisfactory solution is obtained.

Figure 1.1 shows a model of this process of iteration.

CHAPTER 2. GENERAL CLASSIFICATION OF PUMP TYPE

The first problem facing a plant designer when charged with the responsibility for the specification, selection, and procurement of pumps is to make a preliminary decision on the general type of pump (i.e., positive displacement reciprocating, positive displacement rotary, or centrifugal) which will fulfil each of a series of stated process duties. This preliminary information is necessary in order to:

(a) enter the more detailed decision sub-systems for the different classifications of pumps set down in Parts Two to Six;

(b) start the preparation of documentation in order to complete the first stage of procurement which is to send out enquiries to potential suppliers.

The procurement process (b) is time consuming and costs money. It is impractical to send enquiries for every pump to all pump suppliers. It would be too costly and too time consuming. Nor would it be sensible to send all enquiries for every pump to one supplier, only to find that he cannot meet some of the duties, cannot supply others in the time required, or that pumps are much more expensive than forecast. Time is a vital element in the design process for capital projects and all too often the start/completion of detailed pipework design, which is invariably on the critical path for the project, awaits a decision on the choice of pump supplier. User organizations normally have a list of preferred suppliers selected on the basis of experience of their abilities to meet quality, time, and cost criteria. The required pump type is compared with the range of pumps offered by the preferred suppliers and enquiries sent out accordingly. Enquiries are not normally sent out to more than, say, six competing suppliers for particular pumps.

2.1 Primary considerations

The first stage in the decision making process is modelled in Fig. 2.1, Flowchart for general classification of pump type. Based on the process duty sheet and a preliminary layout, the differential static head is calculated and the first estimate of pump head made. From the system requirements a preliminary choice of the number of pumps, and the first estimate of pump flow for each individual pump is made. Flow regulation requirements are considered next. For accurate measurement of small flows over the full range (see section 2.5 for definition of small flows) a positive displacement reciprocating metering pump would be the first choice and the designer should move directly to Part Two for further detailed analysis. For high pressures and small flows, without any specific requirement for accurate measurement of flow, a general purpose reciprocating positive displacement pump would be the first choice and the designer should move to Part Three for further detailed analysis. The effects of fluid properties are then considered and if viscosity is high (see section 2.6 for definition of high viscosity) a positive displacement rotary

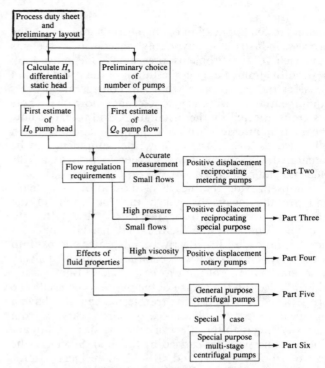

Fig. 2.1. Flowchart for general classification of pump type

pump may be the appropriate choice and the designer should turn to Part Four for further detailed analysis. For all other applications the appropriate choice would be in the general classification of centrifugal pumps and the designer should use Part Five for further detailed analysis, except in the special case of pumps for high pressure duties when the designer should use Part Six. As Part Six is a series of additions or qualifications to Part Five for the special case of pumps for high pressure duties it needs to be read in conjunction with Part Five.

The design process for the selection of plant and equipment for the process industries is also a very complex one. How are decisions made on the time needed or the design man-hours required to select a particular item of plant or to select a group of similar or different plant items? The answer is often in the same category as 'It's as long as a piece of string. . .'. Within broad limits the time fixed for the completion of a project will fix the design time available for the selection of equipment and the staff currently available often fixes the design effort applied to this process. Experience over a long period of time will influence decisions on the levels of staff available and the time allocated to complete specific projects. The degree of risk involved in the selection and purchasing decisions involved in this process is then difficult to assess and must, of needs, be on a rather random basis.

Conversely, not all pumping installations require the same degree of sophistication in the approach to selection. Clearly pumps required for duties involving high temperature, high pressure, handling fluids of high viscosity, or with suspended solids require a much more

4

sophisticated treatment than, say, a pump for a duty involving cold, clean water pumping from a large reservoir with a high inlet static head to a low pressure outlet.

How does an engineer decide on the level of sophistication to apply to a particular pump selection problem? How does he decide on the amount of time to allocate to the problem? In some large process industry companies a more pro-active rational approach to the design process for the selection of equipment is attempted. The detailed approach varies but the principles are essentially the same, i.e., the degree of risk involved in each installation is categorized by assessing the effects of failure on a number of factors. These would typically include the following:

availability of standby plant (including levels of major spares);
financial consequences;
safety of personnel;
environmental considerations;
fluid hazard considerations;
maturity of equipment design concepts and manufacturing techniques (e.g., a well-proven design is likely to involve less risk than an innovative design);
complexity of equipment manufacturing processes.

Each of these factors would be assessed and given a rating with a higher rate for higher levels of risk. The total rate for all the factors would then be used to assess the level of sophistication and quality assurance applied to the selection of particular items of plant. Clearly, the operation of such a system requires experienced (mature) judgement, but it would not be too difficult for an engineer to develop a simple system applicable to his own situation based upon the above (or similar) factors.

The approach adopted here for the selection of pumps for process duties is to look at how the pumps can be integrated into the process systems in order to obtain safe, efficient, and reliable operation over long periods of time. The process of selection is broken down into a series of decision sub-systems starting with a preliminary selection of pump type, followed by iterative checks on flow/head rating, inlet conditions, power rating, casing pressure rating, and sealing considerations, where appropriate. For each part of the publication, dealing with a general classification of pump type, the whole decision-making process is modelled in a flowchart (see Figs 3.1, 9.1, 15.1 and 22.1).

Each of these flowcharts breaks down into three stages.

(a) Stage 1 deals with the preliminary choice of pump type and speed, and is essentially the first level of iteration.
(b) Stage 2 deals with a series of checks and corrections to the first choice for a series of factors and comprises a number of levels of further iteration leading to a confirmation of the pump type and speed.

(c) Stage 3 deals with all those items dependent on the choice of pump type and speed which are necessary to complete the pump specification for selection purposes.

For pumps in relatively simple installations with a low level of risk the plant designer could use only Stage 1, miss out Stage 2, and go directly to Stage 3. Only for more complex installations with higher levels of risk would the full three stage process be applied. With greater experience, the plant designer could use only selected checks from Stage 2.

The process of selection is then broken down into a series of decision subsystems starting with a preliminary selection of pump type and speed, followed by iterative checks on flow/head rating, inlet conditions, power rating, casing pressure rating, and sealing considerations, where appropriate. Each chapter of the four main parts of the publication deals with a separate decision sub-system. In each chapter the sub-system of checks and calculations is first modelled in the form of a flowchart so that the reader can easily chart his way through the complex process of iteration and see how each part of this process relates to the whole and to other parts of the system. The flowchart models are then followed by detailed guidelines on how each stage of the checks and calculations is carried out. These guidelines are based upon a series of empirically derived formulae and criteria.

This methodology provides a series of short cuts to the practicing engineer who, as he gains more experience, can adopt a 'cafeteria' approach and pick and choose from the 'menu' of parameters that need either a simple or rigorous approach depending on the process, the fluids, and the circumstances of the installation. The flowchart models of Figs 3.1, 9.1, 15.1, and 22.1 provide a framework within which the engineer's experience can be catalogued and analysed to result in a greater depth of understanding of the processes involved.

2.2 Preliminary choice of number of pumps

For most current designs of process pumps, whether positive displacement reciprocating, positive displacement rotary, or centrifugal, the L10 life is less than 8000 hours. Consequently, pumps fall into reliability classes 4, 5, or 6 as defined in Appendix II, and the typical arrangement is to have one running pump rated at 100 per cent duty, with one identical pump as the standby (either installed or available as a spare). Other considerations are:

(a) systems that cannot tolerate an interval of no-flow upon changeover to a standby pump, require two running pumps each rated at 50 per cent duty flow together with a third identical pump as an installed spare;
(b) systems which require constant pressure should be assumed to include an accumulator. Schemes with two continuously running pumps each rated at 100 per cent duty should *not* be the first choice as this leads to high running costs.

2.3 First estimate of pump flow

Normal flow is taken as the largest process flow required for operation at the rated plant output. Estimate the corresponding flow for one pump (Q_0) using the number of pumps required for the duty and system requirements from 2.2. Further analysis of pump flow will be required in Parts Two to Five.

2.4 First estimate of pump head

First estimate the pump head as

$$H_0 = H_s + H_r \quad \text{(m)}$$

where H_r is the total frictional head loss in the inlet and delivery system at normal flow, and H_s is the differential static head across the pump, given by

$$H_s = (h_{sd} - h_{si}) + \frac{10.2(P_d - P_i)}{\rho} \quad \text{(m)}$$

where

h_{sd} = static liquid head from pump centre to delivery vessel free liquid surface or constant pressure point (m)

h_{si} = static liquid head from pump centre to inlet vessel free liquid surface (m)

  (Note that h_{sd} and h_{si} are negative when the appropriate liquid surface lies below the pump centre.)

P_d = gas pressure at free surface of the liquid in the delivery vessel, or the pressure at the constant pressure point in the delivery system (bar g)

P_i = gas pressure at the free surface of the liquid in the inlet vessel (bar g)

  (Note that P_d and P_i are negative for vessels under vacuum.)

ρ = liquid density (kg/l)

  For slurries the density of the mixture must be used.

Note

All systems can be reduced to a transfer of liquid between two reservoirs or to circulation to and from one reservoir.

The steady pressure in each reservoir (P_d or P_i) can be either a gas pressure on the free surface of the liquid or a point in the liquid system deliberately maintained at constant pressure by automatic control.

The static liquid head (h_s) is defined as the difference in elevation between the pump centre and either the free surface of the liquid or the constant pressure point in the liquid system.

Especial note must be taken of the fact that the data from which pump duties are determined are always subject to uncertainty.

The process flowsheet lays down the required liquid flowrates and the gas pressures in vessels. The latter are usually maintained at fixed values by instrumented control systems. However, the liquid heads due to elevation and the pressure losses due to friction in the piping system depend upon the dimensions of the associated vessels and upon the plant layout.

Experience shows that elevations are estimated with reasonable precision, viz. within ± 10 per cent, but that estimates of frictional losses are much less precise. Fortunately for typical process plant duties, the effect of this uncertainty in the estimates of frictional losses is often masked by the relatively larger heads needed to overcome the differentials in elevation and gas pressure. For those duties where the frictional losses are dominant, a review at the latest data consistent with the project programme is advisable, remembering that:

(a) the ordinary correlation between friction loss and Reynolds number is no better than ± 20 per cent at the 95 per cent confidence level;

(b) losses in pipework fittings or components like heat exchangers can be subject to large errors when synthesized from geometrical data (Ideally such values should be obtained empirically from compatible units in similar service.);

(c) the greatest source of error may lie simply in under-estimating the complexity of the piping system at the beginning of the design stage (This is especially true of hot systems where the thrusts and moments exerted by pipework during thermal expansion are constrained to very low values.).

2.5 Considerations of flow measurement

For

$$\frac{\rho \cdot Q_0}{H_0} < 0.8 \quad \text{and} \quad Q < 0.8 \text{ l/s}$$

First choose a positive displacement reciprocating metering pump, and move to Part Two.

For flows exceeding 0.4 l/s at pressures exceeding 60 bar g where accurate measurement is not a requirement, first choose a special purpose reciprocating positive displacement pump, and move to Part Three.

For rough measurement of small flows, the speed of a variable-speed rotary pump gives an indication of flow provided that the liquid is clean and its viscosity is sufficiently high, viz.

$$v > 1.6\left(\frac{H_0}{Q_0}\right)$$

where v is the kinematic viscosity in cSt.

To meet this requirement the designer should first select a rotary positive displacement pump and move to Part Four.

2.6   Consideration of the effect of viscosity

Calculate

$$\phi = \frac{v}{Q_0^{1/2} \cdot H_0^{1/4}}$$

where v = kinematic viscosity of the liquid at normal temperature (cSt).

For $\phi > 20$ a positive displacement rotary pump should be selected and the designer should move to Part Four.

For $\phi < 20$ a centrifugal pump should be selected and the designer should move to Part Five.

For definition of ϕ see section 23.5.

2.7   High head duties

If $H > 470Q^{1/2}$ ($Q < 40$) first consider a reciprocating pump and move to Part Three.

For pump capacities above 1.6 l/s see also Part Four.

PART TWO

Positive displacement pumps: reciprocating metering

CHAPTER 3. INTRODUCTION

The system of calculations provided will allow the reader to develop the specification of the pump duty for enquiries to be sent out to pump vendors, using the Pump Data Sheet (Appendix IX), and provide an estimation of the characteristics and requirements of the pump(s) in order that related design work (i.e., civil, piping, electrics, instruments, etc.) can proceed in parallel.

Metering pumps account for a relatively small proportion of all pumps used on process plants. However, with the increasing trend towards more automation of process plants the proportion of metering pumps must inevitably increase as the measurement of flow to fine tolerances by any other means is notoriously difficult. The positive displacement reciprocating metering pump is also a fairly recent genre, the first unit being developed in the United States in 1936. Apart from the direct measurements of small flows, one of the main applications is for the injection of measured small quantities of fluid into high pressure process streams, and they are used universally across the whole range of process industries. Fluids handled range from viscous slurries, to solvents, corrosive fluids, radioactive fluids, and liquids that must be maintained at elevated temperatures.

For ease of treatment the text has been grouped into chapters dealing with the preliminary choice of pump, inlet conditions, flow/head rating sequence, driver power rating, casing pressure rating, and sealing considerations. Each chapter is supported by a flow diagram providing a model of the sub-system of calculations. The tacit assumption underlying this arrangement is that the same treatment will be applied to each and every pump installation.

The flow diagram shown at Fig. 3.1 provides a model of the whole system of calculations to be carried out for reciprocating metering pumps. Each operation is cross-referenced to the relevant chapters or sections as appropriate (numbers in right hand side of boxes). As outlined in Part One, the system is broken down into three stages in order to cope with the needs of different users and/or different pump installations. The first stage covers the preliminary choice of pump type and speed; the second stage covers the more sophisticated treatment with particular reference to metering criteria, corrections to the NPSH available, and fluid properties; and the third stage covers information requirements for completion of the data sheet and for the needs of other related design work, (i.e., civil, piping, electrical, instruments, etc.) which are dependent on the choice of pump type and speed.

Metering pumps are usually procured on the basis of manufacturers' technical specification for their mechanical design and construction; the purchase decision would then depend only on the capital cost of the pump. The treatment given here is intended to guide the buyer in ensuring that the vendors' offers are on the same technical footing.

For some projects, such as plant extensions, the simplest course of action is to use pumps that have already proven to be satisfactory for the existing plant. In this case it may be unnecessary to proceed further into the design process than Stage 1 of the flow diagram of Fig. 3.1. If the new duties are sufficiently different from the existing pump duties, it may be necessary to proceed further and consider some of the factors of Stage 2.

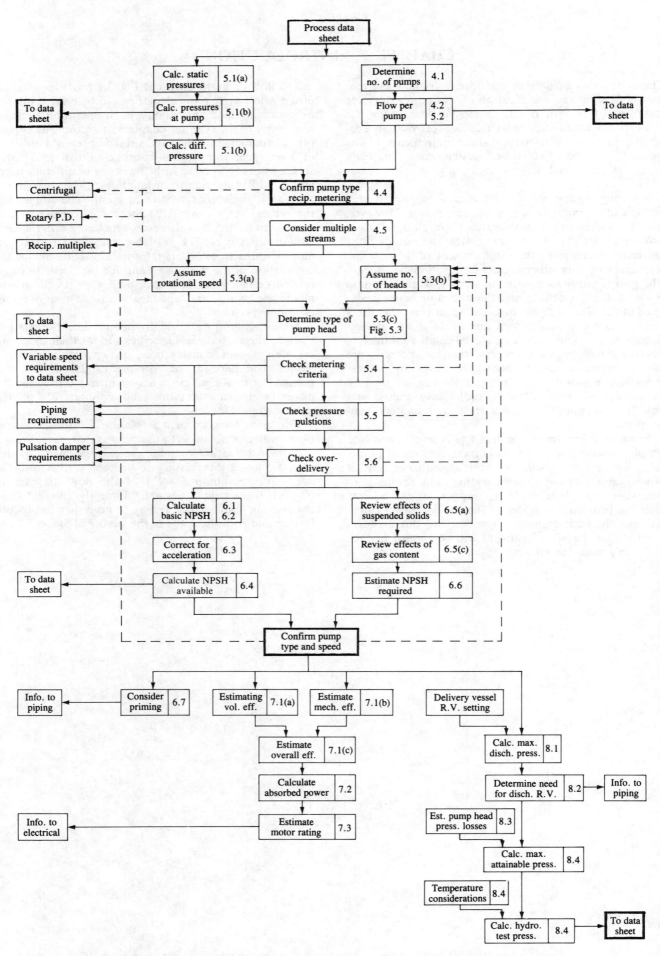

Fig. 3.1. Flowchart of the decision-making process for the selection of reciprocating metering pumps

CHAPTER 4.   PRELIMINARY CHOICE OF PUMP

The system used for the preliminary choice of metering pumps is modelled in Fig. 4.1.

4.1   Preliminary choice of number of pumps

Reciprocating metering pumps fall into reliability classes 4, 5, or 6 as defined in Appendix II. The standard arrangement is one running pump rated at 100 per cent duty, with one identical pump as the spare either installed or available as a replacement.

4.2   Estimate of pump flow

Normal flow is taken as the largest process flow required for operation at the rated plant daily output.

Metering duties are seldom known accurately. Unless the process duty specifically limits the maximum flow, take 110 per cent of this normal flow through to obtain the pump flow rate Q_0 in l/s.

4.3   Estimate of pump mean differential pressure

Estimate the maximum mean differential pressure P_0 as section 5.1 for the normal flow.

4.4   Consideration of flow metering requirements

(a)   *Small flows*

If $\left(\dfrac{Q_0 \cdot \rho}{P_0}\right) \leqslant 0.8$ and $Q_0 \leqslant 0.8$

where Q_0 = flow rate (l/s)

ρ = liquid density (kg/l)

$P_0 = P_d - P_i$ (bar)

P_d = mean discharge pressure (bar g)

P_i = mean inlet pressure (bar g)

(i)   *Accurate flow metering* (*better than ± 10 per cent*)
Use a reciprocating variable stroke metering pump.

Take the flowrange as 0–100 per cent pump flow, unless otherwise specified by process considerations.

(ii)   *Rough flow metering* (*when $0.8 < Q_0 < 8$*)
If a metering accuracy worse than ± 10 per cent pump flowrate is acceptable use a sliding vane rotary pump with variable speed drive if

$$v > \frac{16 P_0}{Q_0}$$

where v = kinematic viscosity of liquid (cSt)

See Part Four.

(iii)   *Batch metering*
Use a reciprocating metering pump with stroke counting or time controlling device.

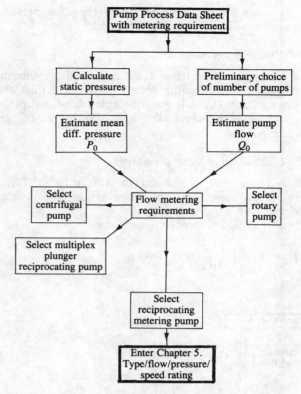

Fig. 4.1.   Preliminary choice of pumps

(b)   *Large flows*

(i)   If $\left(\dfrac{Q_0 \cdot \rho}{P_0}\right) > 0.8$

use a centrifugal pump with a standard flowmetering system. See Part Five.

(ii)   If $\left(\dfrac{Q_0 \cdot \rho}{P_0}\right) < 0.8$ and $Q_0 > 0.8$ l/s

use either
   a multiplex plunger reciprocating pump
or
   a special-purpose centrifugal pump if liquid is clean. See Part Five.

4.5   Multiple stream applications

The combination of variable-stroke metering pump heads for different streams on to a common drive unit should be considered for all multiple stream applications. Proportioning should be effected by stroke adjustment of individual heads with total flow controlled by a variable speed drive.

CHAPTER 5. TYPE/FLOW/PRESSURE/SPEED RATING

The process for the specification of type/flow/pressure and speed rating is modelled in Fig. 5.1.

5.1 Pumping pressures

The pulsating flow from each stroke of a metering pump results in pulsating pressures at pump inlet and discharge. Consequently the differential head and pressure have not the same significance as for centrifugal pumps.

(a) Calculation of static pressures

All systems can be reduced to a transfer of liquid between two reservoirs or to circulation to and from one reservoir.

The inlet static pressure (P_{si}) is given by

$$P_{si} = P_I + \left(\frac{\rho \cdot h_{si}}{10.2}\right) \quad \text{(bar g)}$$

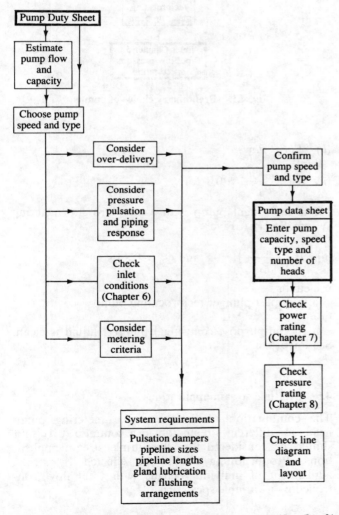

Fig. 5.1. Reciprocating metering pumps – type/flow/head/speed rating sequence

The discharge static pressure (P_{sd}) is given by

$$P_{sd} = P_D + \left(\frac{\rho \cdot h_{sd}}{10.2}\right) \quad \text{(bar g)}$$

The differential static pressure (P_s) is given by

$$P_s = P_{sd} - P_{si} \quad \text{(bar)}$$

where

h_{sd} = static liquid head from pump centre to delivery vessel free liquid surface or constant pressure point (m)

h_{si} = static liquid head from pump centre to inlet vessel free liquid surface (m)

(Note that h_{sd} and h_{si} are negative when the appropriate liquid surface lies below the pump centre).

P_D = gas pressure at free surface of the liquid in the delivery vessel, or the pressure at the constant pressure point in the delivery system (bar g)

P_I = gas pressure at the free surface of the liquid in the inlet vessel (bar g)

(Note that P_D and P_I are negative for vessels under vacuum.)

ρ = liquid density (kg/l)

Check the proposed layout to see if the piping system rises above the delivery vessel to form a syphon. If it does then calculate the maximum discharge static pressure P_{sd} with the static head h_{sd} taken to the point of highest elevation of the piping system. There should be no syphons on the inlet piping system.

(b) Calculation of duty pressures

The duty pressures should be taken as

$$P_i = P_{si} - \left(\frac{\rho \cdot h_{ri}}{10.2}\right) \quad \text{(bar g)}$$

$$P_d = P_{sd} + \left(\frac{\rho \cdot h_{rd}}{10.2}\right) \quad \text{(bar g)}$$

$$P_0 = P_d - P_i \quad \text{(bar)}$$

where

P_i = mean inlet pressure (bar g)

P_d = mean discharge pressure (bar g)

h_{ri} = frictional head loss in inlet piping system at mean flow (m)

h_{rd} = frictional head loss in delivery piping system at mean flow (m)

P_0 = mean differential pressure (bar)

Check the maximum and minimum values of P_i, P_d, and P_0 for which the pump should still operate. Self

14

consistent sets of conditions should be given on the Process Data Sheet.

5.2 Pump flowrate and capacity

Take the pump flow, Q_0 from section 4.2. Note that as metering pumps are manufactured in a limited number of plunger diameters, stroke lengths, and speeds, the actual pump capacity, Q_p, is likely to exceed the required pump flow, Q_0.

5.3 Guide to pump speed and type

(a) Rotational speed

The rotational speed of metering pumps is limited by the following considerations:

  suppression of gas release;
  unidirectional loading of motion work;
  avoidance of pressure wave enhancement;
  viscous losses affecting volumetric efficiency.

As a first estimate take the crank rotational speed, N, as 1 rev/s for low viscosity liquids or the lower speeds shown in Fig. 5.2 for pumps handling viscous liquids.

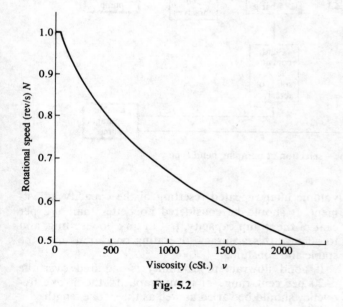

Fig. 5.2

Enter section 5.3(c) with Q_e where

$$Q_e = \left(\frac{Q_0}{N}\right)$$

(b) Pump arrangement

The preliminary choice is a simplex pump without pulsation dampers.

(c) Pumping head type (see Fig. 5.3)

Pump heads of the plunger type should be used for all duties except where:

  (i)   slight gland leakage cannot be tolerated – because of sterility, toxicity, corrosion, or the high cost of the lost fluid;
  (ii) the process fluid contains abrasive particles or precipitates in gland and an alternative flushing medium compatible with the pumped fluid and acceptable to the process is not available (Note that the gland flush reduces metering accuracy.);
  (iii) a lubricating medium compatible with the pumped fluid and acceptable to the process is not available;
  (iv) for high pressure and low flows where loss of metering accuracy due to gland leakage and hold-up is unacceptable. As a guide take loss of flowrate repeatability due to glands as

$$\left(\frac{N \cdot Z}{10 \cdot Q_0}\right)^{2/3} \log P_d; \quad P_d > 12 \ (\%)$$

where Z = number of plungers
      (take as 1 for first estimate)

Diaphragm pump heads should be used for these other duties.

  (i)   *Mechanically actuated metal diaphragm*
        Use if $Q_e < 0.02$ l/s and $P_d < 5$ bar g and leakage on diaphragm failure is acceptable.
        Note
        PTFE or elastomer diaphragms and bellows are available for flows to 0.8 l/s but are only suitable where short diaphragm lives (typically 1000 hours) and low accuracy (± 10 per cent) are acceptable.
  (ii) *Hydraulically actuated single diaphragm*
        Use on duties other than the above. If the process fluid contains large solids or abrasive particles a design without a support plate in the process fluid should be specified. This may be either a single or double diaphragm head.
  (iii) *Double diaphragm*
        Use if cross contamination of process and hydraulic fluids on diaphragm failure is unacceptable.

(d) Confirm selection

Confirm selection of pump type and speed by considering the following.

  (i)   Metering criteria (section 5.4).
  (ii)   Pressure pulsation (section 5.5).
  (iii) Over-delivery (section 5.6).
  (iv)   Inlet conditions (Chapter 6).
  (v)   Process data for the liquid concerned.
  (vi)   Catalogues of preferred vendors.
  (vii) Plant experience from similar installations.

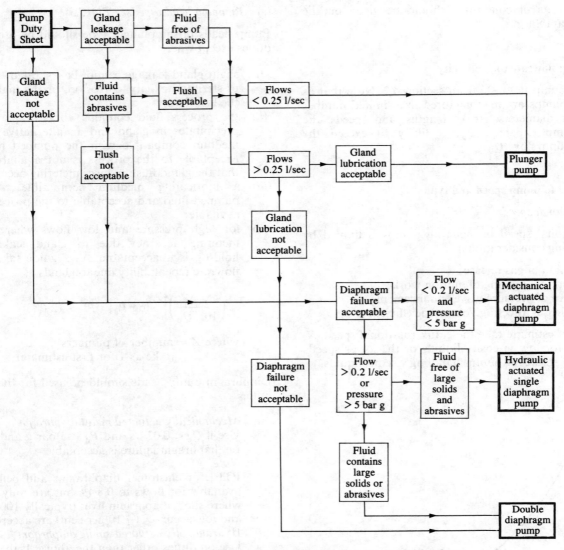

Fig. 5.3.   Reciprocating metering pumps – selection of pumping head type

5.4   Metering criteria

The accuracy of metering is not solely determined by the accuracy of pump manufacture but also by factors connected with the installation such as fluid, pressures, temperatures, contamination, and electrical supply frequency and voltage. Although extremely high accuracies (errors of the order of ±0.05 per cent) can be achieved under favourable conditions, accuracy in practical installations should be considered no better than ±2.5 per cent. To achieve this accuracy the fluid must be clean and operating conditions stable and constant. Slurries or widely ranging conditions, particularly with high pressures, preclude accurate metering.

(a)   Flowrate repeatability

Flowrate repeatability is defined as the ability of a metering pump to deliver consistently a measured volume after repeated resetting of the capacity adjustment. It should be considered no better than ±2 per cent of the pump capacity, Q_p, at any flow setting, and to achieve this accuracy, operating conditions must be stable and constant.

If good flowrate repeatability is required over the 0–20 per cent range of pump capacity the stroke frequency should be varied as well as the stroke length.

Wide ranges of operating pressures, temperatures, and fluid properties will result in a considerably reduced repeatability. See section 7.1(a) for the change in volumetric efficiency with discharge pressure.

(b)   Accuracy of mean flow measurement

Although the flowrate corresponding to a given setting value may be repeatable, the absolute value of this flowrate is subject to a further tolerance of the order of 0.5 per cent of the pump capacity.

(c)   Flowrate continuity

The instantaneous flowrate from a metering pump may fluctuate markedly over each crankshaft revolution. See section 6.2 for prediction of the peak instantaneous flow: mean flow. If duties require continuous delivery flow, install a discharge pulsation damper or specify a multiplex pump.

For smoothing delivery flow with a damper, the delivery system should be damped by resistance. Typically, for a tolerance on flow of the same order as the flowrate repeatability, ensure, for low viscosity liquids

$$h_{rd} > \left(\frac{5.1 \, P_r^{0.7}}{\rho} \right) \quad \text{and} \quad h_{rd} > \left(\frac{1.3 \, P_{rd}}{\rho} \right)$$

The flow of viscous liquids is more effectively smoothed by using multiplex pumps.

Dampers necessarily have a long time constant which limits the rate at which the desired flow setting under automatic control can be changed and also requires a substantially constant discharge pressure. For a tolerance on flow continuity of the same order as the flowrate repeatability this limit on rate of flow setting gives a minimum of

$$\left(\frac{500}{Z} \right)$$

times the period of one crankshaft revolution for full travel of the flow setting (from 0 to 100 per cent), where Z is the number of plungers.

5.5   Pressure pulsation

To avoid pulsation enhancement, the length of piping between the pump ports and a large discontinuity such as a vessel should be less than 4 per cent of the fundamental wavelength (Λ) where

$$\Lambda = \left(\frac{C}{N \cdot Z} \right)$$

where

    C = velocity of sound in fluid (m/s)
    N = crank rotational speed (rev/s)
    Z = number of plungers.

Then calculate the pressure pulsation at both inlet and discharge of pump from

$$\delta P = \frac{700\rho \cdot N \cdot Q}{Z} \sum \left(\frac{1}{d^2} \right)$$

where

    δP = pressure pulsation (peak to peak) (bar)
    l = length of line to large discontinuity (m)
    d = diameter of a section of line of length l

Alternatively, if the system presents no discontinuity in the piping close to the pump, then calculate the pressure pulsation at inlet and discharge of pump from

$$\delta P = \frac{K \cdot \rho \cdot C \cdot Q}{D^2}$$

where

    D = diameter of line bore (mm)
    K = 40 for simplex pump
         20 for duplex pump
          3 for triplex pump
          1 for quintuplex pump

For lines of varying diameter, take D as the minimum line diameter excluding short lengths ($< 0.01\Lambda$).
If

$$\delta P > 0.01P$$

and

$$\delta P > 0.04P^{0.7}$$

where

    P = mean pressure at pump inlet and discharge
                               (bar a)

then the pressure pulsations are unacceptable and piping vibration is likely. Reduce the level by:

(a)   increasing line diameter;
(b)   specifying pulsation damper;
(c)   specifying multi-headed pump.

5.6   Over-delivery

The pump discharge pressure should always exceed the inlet pressure to avoid flow through the pump which is not controlled by the plunger.
For simplex pumps without pulsation dampers ensure that

$$P_s > \frac{\delta P_i}{2}$$

and

$$P_s > \frac{\delta P_d}{2}$$

For other arrangements ensure that

$$P_s > \left(\frac{\delta P_i}{2} + \frac{\delta P_d}{2} \right)$$

where

    δP_i = pressure pulsation at pump inlet (bar)
          (from section 5.5)
    δP_d = pressure pulsation at pump discharge (bar)
          (from section 5.5)

Pressure retaining valves or spring loaded pump valves to develop an artificial pressure difference should be avoided.

On batch systems, drips from the delivery line should be prevented by installing valves at the outlet or by arranging for the outlet to be at the highest point of the pipeline and of small bore.

CHAPTER 6.   INLET CONDITIONS

The process of confirming that the inlet conditions are suitable for the preliminary choice of pump type and speed is modelled in Fig. 6.1.

6.1   Calculation of basic NPSH

$$\text{Basic NPSH} = h_{si} + \frac{10.2}{\rho}\left(\frac{B}{1000} + P_1 - P_v\right) \quad \text{(m)}$$

where

h_{si} = static liquid head over pump inlet (m)

This is measured from the lowest liquid level to the pump centre and is negative when the liquid level is below the pump centre. The lowest level is not necessarily the extreme value but is the one at which the pump must still operate.

B = minimum barometric pressure at pump location (mbar)

Use 0.94 of the mean barometric pressure from local meteorological data to allow for weather changes.

P_1 = minimum working value of the gas pressure on the free liquid surface in the inlet vessel. This is negative for vessels under vacuum (bar g)

P_v = vapour pressure of the liquid at the maximum operating temperature (bar a)

When the fluid is complex – generally a light hydrocarbon – the 'bubble point' pressure should be used, not the Reid vapour pressure nor pressure readings from the vapour space of the supply vessel.

ρ = liquid density at the maximum operating temperature. For slurries the density of the *mixture* should be used (kg/l)

6.2   Correction for frictional head

Calculate the maximum frictional head loss in the inlet pipe system, H_{ri} at peak instantaneous flow rather than at mean flow and include the maximum operating loss in the inlet filter when this is dirty.

Obtain the peak instantaneous flow from Table 6.1.

Table 6.1

Configuration	$\dfrac{\textit{Peak flow}}{\textit{Mean flow}}$
Quintuplex	1.03
Triplex	1.1
Duplex	1.6
Simplex	3.2

6.3   Correction for acceleration head

Calculate H_{ai} the head required to accelerate the fluid in the inlet piping during each pulsation cycle from

$$H_{ai} = \left(\frac{6\delta P_i}{\rho}\right) \quad \text{(m)}$$

where

δP_i = pressure pulsation at pump inlet (bar) (from section 5.5)

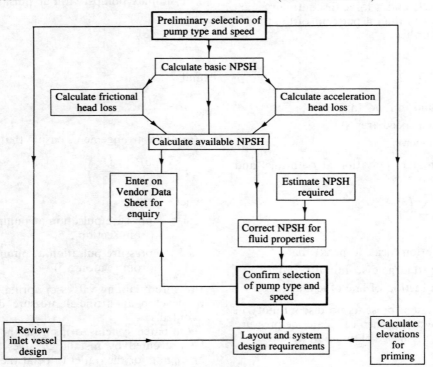

Fig. 6.1.   Inlet conditions

18

6.4 Calculation of available NPSH

Calculate the available NPSH at the pump suction flange from:

(a) for simplex and duplex pumps

$$\text{Available NPSH} = \text{Basic NPSH} - (H_{ai}^2 + H_{ri}^2)^{1/2} - 1 \quad \text{(m)}$$

(b) for triplex and quintuplex pumps

$$\text{Available NPSH} = \text{Basic NPSH} - (H_{ai} + H_{ri}) - 1 \quad \text{(m)}$$

6.5 Effect of fluid properties on NPSH

The following corrections are to be used for discussions on layout and system design. They would not normally be disclosed to the pump supplier.

(a) Suspended solids content

The behaviour with slurries is notoriously difficult to predict; the only reliable test is to run the pump on the plant. Pastes, gels, and other non-Newtonian fluids must always be the subject of experimental tests. For suspensions of hard crystalline solids, use the following empirical rules as a guide:

(i) very fine suspensions where all the particles are less than 3 μm are best treated as viscous liquids – see Fig. 6.2;

(ii) for slurries having a range of particle sizes take

$$\text{NPSH}_{\text{slurry}} = \left(\frac{1}{1+x}\right) \cdot \text{NPSH}_{\text{carrier}}$$

where

x = fractional content by volume of solid particles and the NPSH for the carrier liquid has already been corrected for viscosity.

(b) Viscosity

An increase in viscosity increases the NPSH required. For high values of viscosity the losses across valves are approximately proportional to the liquid velocity. Consequently the NPSH required is directly proportional to viscosity at constant speed and flow.

However, it is normal practice to reduce pump speed as viscosity increases (see section 5.3(a)) and so, in effect, limit the variation of NPSH required with viscosity change.

(c) Gas content

For accurate metering, the liquid should contain no free gas.

As pressure reduces across the pump inlet valves on the inlet stroke, the dissolved gas may be released from solution if the available NPSH is too small. It is essential to avoid gas release for metering duties.

6.6 Estimation of NPSH required

The NPSH required by a pump is dictated by its detail design and its operating speed.

As a guide for layout, system, and pump specification, a preliminary estimate of the NPSH required by plunger and mechanically actuated diaphragm pumps may be obtained from Fig. 6.3 where
For single valves

$$A = \left(\frac{24v \cdot Q \cdot \rho}{Z \cdot d_v^3}\right) + 5.10^5 \left(\frac{\rho \cdot Q^2}{Z^2 \cdot d_v^4}\right)$$

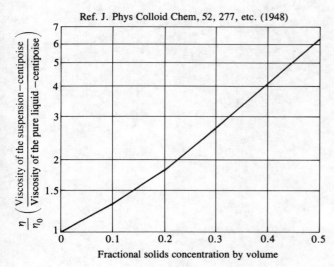

Ref. J. Phys Colloid Chem, 52, 277, etc. (1948)

Fig. 6.2

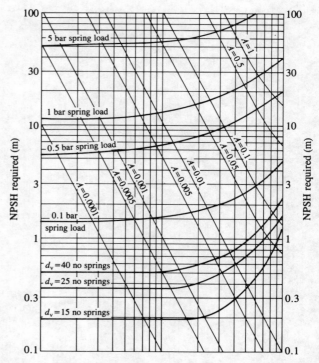

Fig. 6.3. **Estimate of NPSH required as a function of valve spring load**

For double valves

$$A = \left(\frac{80v \cdot Q \cdot \rho}{Z \cdot d_v^3} \right) + 15.10^5 \left(\frac{\rho \cdot Q^2}{Z^2 \cdot d_v^4} \right)$$

where

d_v = nominal valve size (mm)

For hydraulically actuated diaphragm pumps, estimate as above and take additional diaphragm and support plate losses as for a single valve without spring loading.

The actual NPSH required should be checked after selection of the pump supplier.

6.7   Priming

For accurate metering, pumps should have a flooded inlet and the gland subject to a pressure above atmospheric pressure to avoid inward air leakage.

The pressure at the pump inlet should be greater than the pressure differential required to open both inlet and discharge valves against their spring forces.

When the pump does not auto-prime through the process piping but is primed by operation of a vent valve, check the line diagram to ensure that:

(a)   the vent is shown;

(b)   the vent is piped to a safe disposal point when the pump is handling a hazardous liquid;

(c)   when the inlet vessel is working under vacuum, the vent is piped to the gas space of the inlet vessel;

(d)   when the vent is piped, then a visible indicator is fitted in the vent line adjacent to the pump.

(a)   Liquified gases

With liquified gases at sub-ambient temperature, the discharge pressure at start-up should be less than the pressure reached if a cylinder fills with vapour and the plunger acts as a compressor. As the clearance volume is likely to be of the same order as the swept volume, the maximum absolute pressure developed in the cylinder is unlikely to exceed twice the absolute inlet pressure. The discharge pressure should be less than this value to clear the cylinder of vapour. Note that vapour compression does not result in condensation because it is nearly adiabatic and hence yields superheated vapour.

This condition normally requires a special bypass provision. Check that this has been included in the engineering line diagram.

Metering cannot begin until the pump has reached its working temperature; the inlet lines should be lagged to achieve this quickly.

CHAPTER 7.   POWER RATING

The power rating sequence is modelled in Fig. 7.1.

7.1   Pump efficiency

(a)   *Volumetric efficiency*

Volumetric efficiency

$$\eta_v = \frac{\text{Delivered volume} \cdot 100}{\text{Swept volume}} \quad (\%)$$

For preliminary estimation of the volumetric efficiency refer to Fig. 7.2. The lower line of each band refers to pumps with small plunger diameters and fluids with high compressibility.

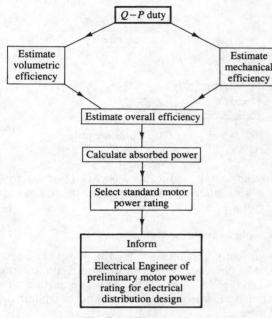

Fig. 7.1.   Power rating

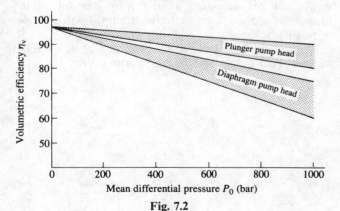

Fig. 7.2

(b)   *Mechanical efficiency*

Mechanical efficiency

$$\eta_m = \frac{\text{Theoretical hydraulic power} \cdot 100}{\text{Power input to pump}} \quad (\%)$$

For preliminary estimation of mechanical efficiency take

$$\eta_m = \frac{100}{\left(1.3 + \dfrac{5}{Q_0 P_0}\right)} \quad (\%)$$

(c)   *Overall efficiency*

The overall pump efficiency (η) is defined as

$$\eta = (\eta_v \cdot \eta_m)/100 \quad (\%)$$

7.2   Calculation of absorbed power

$$E = \left(\frac{10 Q_0 P_0}{\eta}\right) \quad (\text{kW})$$

7.3   Determination of driver power rating

Multiply the mean power requirement for the pump by the following additional factors as appropriate:

belt drive	1.07
simplex pump	1.3
duplex pump	1.1
equipment and data tolerance	1.1

For multiple pump heads driven by one electric motor drive, sum the power requirements for the individual feeds.

For an electric motor drive select the next highest power rating in kW from

0.75	1.1	1.5	2.2	3	4	5.5	7.5	11	15	18.5	22	30	37	45	55

For variable speed drives consult a specialist or the motor manufacturers if necessary for advice on the appropriate motor rating and type.

Note

This power rating should be used only for a *preliminary* estimate of the electrical distribution load. Check the required power rating and frame size after selection of pump using the manufacturers data.

Standard motors have a starting torque of about 150 per cent full load torque which is adequate for reciprocating metering pumps.

CHAPTER 8. CASING PRESSURE RATING

The process for the determination of casing pressure rating is modelled in Fig. 8.1.

8.1 Calculation of maximum discharge pressure

Calculate maximum discharge pressure, P_{dmax}, from

$$P_{dmax} = P_R + \frac{\rho m}{10.2}(H_{sd} + H_{rd}) \qquad \text{(bar g)}$$

where

P_R = set pressure of relief valve on delivery vessel (bar g)

ρ_m = maximum liquid density (kg/l)

H_{sd} = maximum static head from pump to the free liquid surface in the delivery vessel (m)

H_{rd} = maximum frictional head loss in the delivery pipe system at the peak instantaneous flow (see section 6.2 for peak flow: mean flow) (m)

8.2 Discharge pressure relief rating

A discharge pressure limiting device shall be installed unless there is no possibility of a restriction occurring in the discharge system between pump and vessel. Protection by the motor overload trip is not sufficient.

The conventional pressure limiting device is a relief valve in the discharge pipework or in the hydraulic system of diaphragm pump heads. However, relief valves tend to leak after they have operated and detection of a small leak is difficult. Consequently, for metering duties, specify a bursting disc in the discharge pipework or, for a diaphragm pump, the hydraulic relief valve setting at least 125 per cent of the maximum discharge pressure P_{dmax}.

The relief valve should be sized for at least 110 per cent of the pump displacement capacity at the maximum pump speed.

8.3 Calculation of pump head outlet losses

As a preliminary estimate take the pressure increment due to pump head outlet losses, P_h, as twice the press-

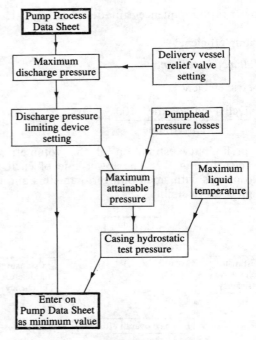

Fig. 8.1. Casing pressure rating

ure corresponding to the inlet NPSH required by the pump. Check after selection of pump using the manufacturers data.

8.4 Casing hydrostatic test pressure

The hydrostatic test pressure for the pump head should exceed 150 per cent of the maximum attainable pressure, taken as the sum of the pressure setting of the discharge pressure limiting device and the pump head outlet losses, P_h.

Pumps handling hot liquid (above 175°C) require individual consideration taking into account piping loads and reduction in material strength with increase in temperature.

The rating of the inlet connection should be identical to that of the casing and discharge connection.

PART THREE

Positive displacement pumps: reciprocating special purpose

CHAPTER 9. INTRODUCTION

Part Three covers the requirements for the specification of the pump duty and selection of special purpose positive displacement reciprocating pumps, i.e., for multiplex, crank driven, reciprocating plunger pumps for delivery pressures exceeding 60 bar absolute and capacities exceeding 0.4 l/sec.

The system of calculations provided will allow the reader to develop the specifications of the pump duty for enquiries to be sent out to pump vendors, using the medium of the Pump Data Sheet (Appendix IX), and provide an estimation of the characteristics and requirements of the pump(s) in order that related design work (i.e., civil, piping, electrics, instruments, etc.) can proceed in parallel.

Reciprocating pumps comprise a very small proportion of all pumps used on process plants (certainly less than 5 per cent). They are particularly suitable for small to medium flows at high pressure, but because of high initial cost and greater maintenance requirements, tend only to be used when no other solution is available. Nevertheless, they perform many critical duties on process plants and there are a wide variety of different configurations to be considered.

For ease of treatment, the text has been grouped into chapters dealing with the preliminary choice of pump, inlet conditions, flow/head rating sequence, driver power rating, casing pressure rating, and sealing considerations. Each chapter is supported by a flow diagram providing a model of the sub-system of calculations. The tacit assumption underlying this arrangement is that the same full sophisticated and rigorous treatment will be applied to each and every pump installation. In practice, however, the treatment applied to the specification and selection of individual pumping installations can vary over a wide spectrum, from a fairly cursory examination of a limited number of critical factors, to the full and rigorous treatment assumed here. As outlined in Part One, this will depend on the critical nature of each installation.

The flow diagram shown at Fig. 9.1, provides a model of the whole system of calculations to be carried out for special purpose reciprocating pumps. Each operation is cross-referenced to the relevant chapters or sections as appropriate (number in right hand side of boxes). As explained in Part One, the system is broken down into three stages in order to cope with the needs of different users and/or different pump installations. The first stage covers the preliminary choice of pump type and speed; the second stage covers the more sophisticated treatment with particular reference to the number of plungers, corrections to the NPSH available, and fluid properties; and the third stage covers information requirements for completion of the data sheet and for the needs of other related design work (i.e., civil, piping, electrical, instruments, etc.) which are dependent on the choice of pump type and speed.

For a very simple pumping installation such as a small cold water duty, or duties with fluids having similar properties to cold water, with single stream operation (i.e., no parallel operation) an engineer might choose to use only stage one, leave out stage two as inappropriate and rely on the pump manufacturer to supply the dependent information of stage three. This would assume that time was not a critical element for the release of information for layout, civil, piping, and electrical design purposes. Again for a relatively simple pumping installation where the requirements were more critical for other aspects of plant design, an engineer may choose to omit stage two and use only stage one followed up by a determination of the dependent information of stage three. Finally for a more complex installation with fluids other than cold water where NPSH available was much more critical, the full rigorous treatment would be applied, including the iterative confirmation of the selection of pump type and speed of stage two before proceeding to stage three. However, as the number of reciprocating pump installations is likely to be small, the more rigorous approach is recommended.

This methodology provides a series of short cuts to the practicing engineer who, as he gains more experience, can adopt a 'cafeteria' approach and pick and choose from the 'menu' of parameters that need either a simple or rigorous approach depending on the process, the fluids, and the circumstances of the installation. The flowchart model of Fig. 9.1 provides a framework within which the engineer's experience can be catalogued and analysed to result in a greater depth of understanding of the processes involved.

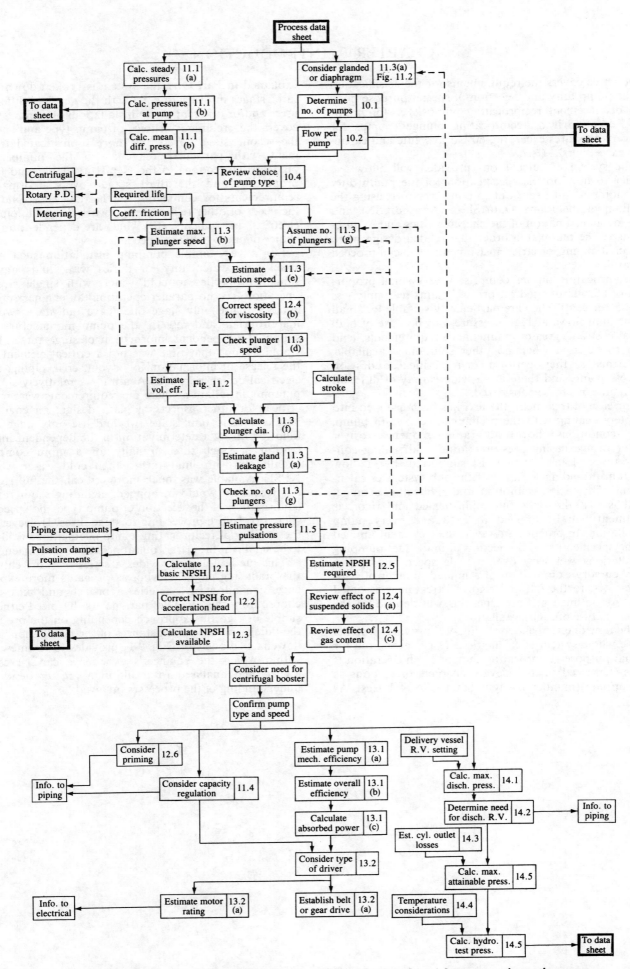

Fig. 9.1. Flowchart of the decision-making process for the selection of special purpose reciprocating pumps

CHAPTER 10.   PRELIMINARY CHOICE OF PUMP

The process for the preliminary choice of pump for reciprocating pumps is modelled in Fig. 10.1.

10.1   Preliminary choice of number of pumps

The reliability classifications are defined in Appendix II.

A crucial decision is whether the process permits the injection of a lubricant. The MTBF of unlubricated glands seldom exceeds 1300 hours, consequently the pumps then fall into Reliability Classes 4, 5, or 6.

Because the MTTR (mean time to repair) usually exceeds 20 hours, the standard arrangements comprise:

(a)   one 100 per cent duty pump with an identical installed spare;

(b)   two 50 per cent duty pumps with an identical installed spare.

The MTBF of lubricated glands can exceed 4000 hours, consequently the pumps can be used in Reliability Class 2 or 3. In principle Class 1 operation is possible.

Other considerations are:

(a)   systems that cannot tolerate an interval of no-flow upon changeover to a standby pump require two running pumps each rated at 50 per cent duty together with a third identical pump as standby;

(b)   systems requiring approximately constant pressure should include an accumulator.

10.2   Pump flow, Q

The total normal flow is taken as the largest process flow required for operation at the rated plant daily output.

Unless the process duty requires a specific maximum flow, take 110 per cent of this normal flow through one pump or divided between pumps as considered in section 10.1, to obtain the pump flow rate Q in l/s.

10.3   Pump mean differential pressure, P

Estimate the maximum mean differential pressure as section 11.1.

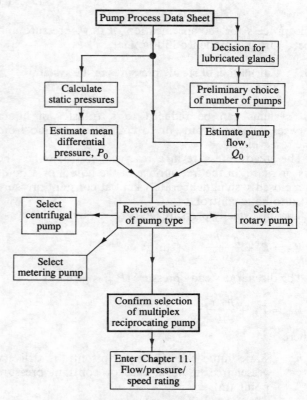

Fig. 10.1.   Preliminary choice of pump

10.4   Review other types of pumps

(a)   If $(Q \cdot \rho)/P > 0.8$ and $Q > 0.8$ (l/s), first consider a centrifugal pump (see Part Five).

(b)   If $v > (16 \cdot P)/Q$ where v is the kinematic viscosity of the liquid at the operating temperature, then first consider a rotary pump (see Part Four).

(c)   If the discharge pressure is less than 60 bar g, or if both $(Q\rho)/P < 0.8$ and $Q < 0.8$ (l/s), then first consider a reciprocating metering pump because of ready commercial availability.

27

CHAPTER 11. SPECIFICATION OF FLOW, PRESSURE, AND SPEED RATING

The process for the specification of flow, pressure and speed rating is modelled in Fig. 11.1.

11.1 Calculation of steady pressures in the system

(a)

All systems can be reduced to a transfer of liquid between two reservoirs, or to circulation to and from one reservoir.

The steady pressure in a reservoir can be either the gas pressure on the free surface of the liquid, or a point in a closed system deliberately kept at constant pressure by automatic control.

The inlet steady pressure (P_{si}) is given by

$$P_{si} = P_I + \frac{\rho \cdot h_{si}}{10.2} \quad \text{(bar g)}$$

The discharge steady pressure (P_{sd}) is given by

$$P_{sd} = P_D + \frac{\rho \cdot h_{sd}}{10.2} \quad \text{(bar g)}$$

where

h_{sd} = static liquid head from pump centre to delivery vessel free liquid surface or constant pressure point (m)

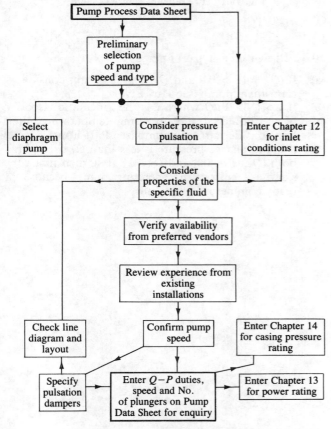

Fig. 11.1. Flow/pressure/speed rating

h_{si} = static liquid head from pump centre to inlet vessel free liquid surface (m)

(Note that h_{sd} and h_{si} are negative when the appropriate liquid surface lies below the pump centre.)

P_D = gas pressure at free surface of the liquid in the delivery vessel, or the pressure at the constant pressure point in the delivery system (bar g)

P_I = gas pressure at the free surface of the liquid in the inlet vessel (bar g)

(Note that P_D and P_I are negative for vessels under vacuum.)

ρ = fluid density, including the effect of suspended solids content (kg/l)

(b) *Calculation of pressures at pump*

The mean pressures at pump inlet and discharge are

$$P_i = P_{si} - \frac{\rho \cdot h_{ri}}{10.2} \quad \text{(bar g)}$$

$$P_d = P_{sd} + \frac{\rho \cdot h_{rd}}{10.2} \quad \text{(bar g)}$$

where

P_i = mean inlet pressure (bar g)

P_d = mean discharge pressure (bar g)

h_{ri} = frictional head loss in inlet piping system at mean flow (m)

h_{rd} = frictional head loss in delivery piping system at mean flow (m)

The mean pressure across the pump is given by

$$P = P_d - P_i \quad \text{(bar)}$$

Check the maximum and minimum values of P_i, P_d, and P for which the pump should still operate. Self-consistent sets of conditions are given on the Process Data Sheet.

11.2 Pump capacity

Take the pump flow, Q from section 10.2.

11.3 Guide to pump type and speed

(a) *Choice of plunger type pump*

Refer to Fig. 11.2 illustrating the main factors in selecting plunger pumps.

Diaphragm pumps should be considered when:

(i) Gland leakage cannot be tolerated on grounds of sterility, toxicity, corrosion, or expense of the loss.

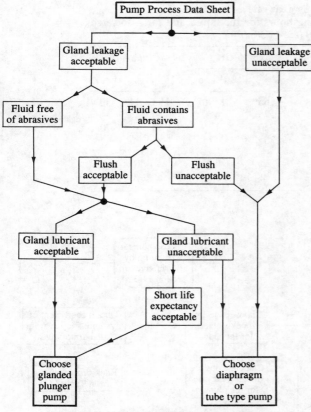

Fig. 11.2. Selection of plunger type pump

The gland leakage rate is insensitive to pressure because the tightening of the gland is proportioned to the pressure. A rough guide to leakage rate is:

$$q = AD \cdot (10 + N) \cdot \log_{10}(P_d + 2P_i + 10)$$

where

D = plunger diameter (mm)

N = crankshaft speed (r/s)

P_d = discharge pressure (bar abs)

P_i = inlet pressure (bar abs)

q = average leakage rate for one plunger (l/hour)

A = constant appropriate to the liquid lubricating the gland, of the order of 3×10^{-4}

(ii) The process fluid contains abrasive particles, or precipitates solids in gland, and a flushing medium compatible with the pumped fluid and acceptable to the process is not available.

(iii) A lubricant compatible with the pumped fluid and acceptable to the process is not available.

(b) *Limit on mean plunger speed set by packing*

The life expectancy of packed glands is difficult to predict: the only satisfactory method is by comparison with a similar pump operating on the same fluid and subject to a similar maintenance regime.

The following criteria enable an estimate to be made of the optimum mean plunger speed, U.

(i) *Plunger wear rate*
Experience shows that the rate of wear is *not* constant. If the desired mean time between plunger changes is T running hours, then

Mean diametrical wear rate

$$= \frac{1.2 \times 10^7}{T^{4/3}} \, \mu\text{m per 1000 hours}$$

(ii) *Effect of pressure*
For a given fluid and a given wear rate

$U \cdot P$ = constant

where

P = differential pressure across the gland (bar)

U = mean plunger speed (m/s)

Use this relationship with caution. Experience suggests that threshold values exist for both U and P, approximately 0.2 m/s for the former and 12 bar for the latter.

(iii) *Effect of lubrication properties of the liquid*
The wear rate correlates with the dynamic coefficient of friction. The form of the correlation has not been determined but an estimate of the wear rate can be made.

First obtain the dynamic coefficient of friction. This requires a laboratory test using a sample of the process liquid.

Next rank the test value for the liquid with those given in Table 11.1. Then enter Table 11.2

Table 11.1. Ranking by coefficient of friction
(Coefficients of dynamic friction between cast iron and alloy steels under boundary lubrication conditions.)

| Rank | Liquid | Viscosity cP @ 20°C | Friction coefficient for plunger material | | |
			1% CrMo	13% Cr	Colmonoy
1	Water	1.0	0.48	0.42	0.21
2	Pentene	0.3	—	0.35	—
3	Adiponitrile	7.0	0.35	0.31	0.23
4	Copper Liquor	2.5	0.24	0.28	0.16
5	98% H_2SO_4	25	0.25	0.20	0.16
6	Mineral Oil	1000	0.25	0.19	0.17
7	Castor Oil	800	—	0.13	—

Table 11.2. Ranking by wear rate

(Ranking of mean wear rates in μm per 1000 running hours for the liquid and for associated values of the product $U \cdot P$ (see 11.3(b)(ii)).)

Rank	Liquid	1% CrMo		13% Cr		Colmonoy	
		$U \cdot P$	Wear	$U \cdot P$	Wear	$U \cdot P$	Wear
−1	Aqueous Caustic soda	—	—	165	1300	165	450
0	Liquid Ammonia	—	—	175	450	—	—
+1	Water	250	600	250	125	250	70
2	Liquid Propylene	205	300	205	25	—	—
4	Copper Liquor	360	100	—	—	—	—
		240	50	—	—	—	—
6	Mineral Oil	100	7	—	—	—	—

Then use Table 11.2 to find the paired values for product $U \cdot P$ and the wear rate. From these the optimum value for U is found.

(c) *Limit to mean plunger speed set by valve operation*

For pumps in Reliability Classes 1 or 2, handling liquids whose working viscosity is less than 12 cSt, the mean liquid velocity through the valves (u) should be limited so that

$$u < \frac{1.5 P_{\mathrm{d}}^{3/8}}{\rho^{1/2}} \quad \text{(m/s)}$$

where

ρ is the liquid density (kg/l)

P_{d} is the discharge pressure (bar a)

The physical constraints on valve size relative to cylinder size produce a relationship between u and U. Take

$$U < \frac{0.11u}{\log N}$$

For viscous liquids reduce both u and N; see section 12.4(b) and Fig. 12.3.

(d) *Estimation of crankshaft rotational speed*, N

Estimate the rotation speed from

$$N = 48\left(\frac{U^3 \cdot Z}{P_{\mathrm{d}} \cdot Q}\right)^{1/2} \quad \text{(rev/s)}$$

where

Q is the volume flow of the pump (l/s)

P_{d} is the discharge pressure (bar abs)

U is the mean piston speed (m/s)

Z is the number of plungers

Implicit in this equation is that the stroke/diameter ratio, S/D, is governed by the needs of the gland according to the relation

$$S/D = 5 \cdot P_{\mathrm{d}}^{1/2} \quad (P_{\mathrm{d}} > 60)$$

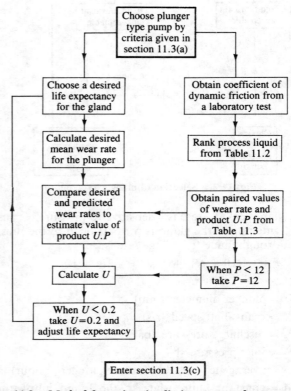

Fig. 11.3. Method for estimating limit on mean plunger speed, U, set by gland wear

Higher rotational speeds require consideration of plunger guidance.

Note also that the capital cost of a pump is strongly influenced by the speed. Consequently competitive pressures drive vendors to offer pumps running at the highest speed that the buyer will accept.

(e) *Calculation of plunger diameter*, D

$$D = \left(\frac{40 \cdot Q}{\eta_{\mathrm{v}} \cdot \pi \cdot S \cdot N \cdot Z}\right)^{1/2} \times 10^4 \quad \text{(mm)}$$

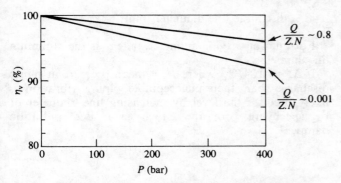

Fig. 11.4. **Volumetric efficiency variation with pressure**

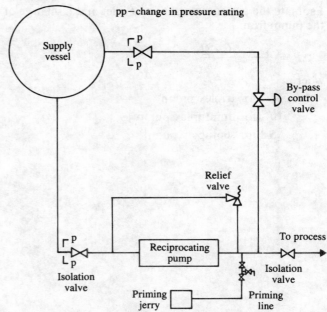

Fig. 11.5. **Line diagram for by-pass control**

The volumetric efficiency η_v is defined by

$$\eta_v = \frac{\text{Discharge volume flowrate}}{\text{Total plunger swept volume rate}} \times 100 \quad (\%)$$

For an estimate of η_v use Fig. 11.4.

Figure 11.4 is based on water, i.e., on a liquid bulk modulus of 2100 MN/m². If the bulk modulus Y of the process liquid is known, then a better estimate, using the value of η_v from Fig. 11.4, is obtained from

$$100 - \frac{Y(100 - \eta_v)}{2100} \quad (\%)$$

See Appendix V for notes on the volumetric properties of liquids.

(f) *Choose the number of plungers based upon the following considerations.*

(i) An odd number is preferred, viz. 3, 5, or 7.
(ii) The motion work bearings limit the plunger load. The convention is to ignore inertia load and consider only the nominal hydraulic load due to the differential pressure. This load is the chief design parameter determining motion work size: the values are chosen by a manufacturer for his range of standard machines.

Pumps with plunger loads exceeding 180 kN are not commercially available; as a rough guide take the allowable load as

$$2.50S^{2.5} \times 10^{-4} \text{ kN} \quad (S > 50)$$

where S is the stroke in mm.

Note that this relation does *not* define a range of similar frames and their motion work, but is the envelope of such ranges.

11.4   Capacity regulation

(a) *Bypass control (Fig. 11.5)*

This is the first choice for systems requiring:

(i) quick response;
(ii) continuous variation down to no-flow.

The disadvantage is that the power absorbed by the pump (at constant pressure) is constant, so that power efficiency falls as the delivery flow decreases.

Check the line diagram to ensure that:

(i) the bypass is returned to the inlet vessel, not the pump inlet line;
(ii) the change in piping and fittings specification occurs at the block valves.

(b) *Speed variation*

This should be considered for:

(i) large pumps, absorbing more than 75 kW;
(ii) pumps handling liquids contaminated by abrasive solids, where control valves may be expected to have a short life and the MTBF of the pump may be extended.

11.5   Pressure pulsation

The fluctuation in volume flowrate produces pressure pulsations in the piping system. The obliquity of the connecting rod introduces a substantial second harmonic of the fundamental pulsation frequency.

To avoid pulsation enhancement, the length of piping between the pump ports and a large discontinuity such as a vessel should be less than 8 per cent of the wavelength for the second harmonic, i.e., less than 4 per cent of the fundamental wavelength λ obtained from

$$\lambda = \frac{C}{N \cdot Z}$$

where

C = velocity of sound in fluid (m/s)

N = crank rotational speed (rev/s)

Z = number of plungers

Estimate the pressure pulsation at inlet and discharge of the pump from

$$\Delta P = \frac{K \cdot \rho \cdot C \cdot Q}{d^2}$$

where

$K = 32$ for triplex pump

$ 10$ for quintuplex pump

$ 5.2$ for septuplex pump

$d = $ discharge line diameter (mm)

For lines of varying diameter, take d as the minimum line diameter.

If $\Delta P > 0.04 P_p^{0.7}$ where $P_p = $ mean pressure at pump discharge, bar, then unacceptable piping vibration is likely. Reduce the level by increasing line diameter or by specifying provision of a gas-loaded pulsation damper.

CHAPTER 12. INLET CONDITIONS

The process of confirming that the inlet conditions are suitable for the preliminary choice of pump is modelled in Fig. 12.1.

12.1 Calculation of basic NPSH

Basic NPSH

$$= h_{si} + \frac{10.2}{\rho} \left(\frac{B}{1000} + P_I - P_v \right) - h_{ri} \qquad \text{(m)}$$

where

h_{si} = static liquid head over pump inlet (m)

This is measured from the lowest liquid level to the pump centre and is negative when the liquid level is below the pump centre. The lowest level is not necessarily the extreme value but is the one at which the pump is still operational.

B = minimum barometric pressure at pump location (mbar)

Use 0.94 of the mean barometric pressure from local meteorological data to allow for weather extreme.

P_I = minimum working value of the gas pressure on the free liquid surface in the inlet vessel. This is negative for vessels under vacuum. (bar g)

P_v = vapour pressure of the liquid at the maximum operating temperature (bar a)

When the fluid is complex – generally a light hydrocarbon – the 'bubble point' pressure should be used, not the Reid vapour pressure or pressure readings from the vapour space of the supply vessel.

ρ = liquid density at the maximum operating temperature (For slurries the density of the *mixture* should be used). (kg/l)

h_{ri} = frictional head loss calculated at the peak instantaneous flow rather than at mean flow, including the maximum operating loss in the inlet filter when this is dirty.

Obtain the peak instantaneous flow from the following Table:

Number of plungers	Peak flow / Mean flow
3	1.115
5	1.035
7	1.018

12.2 Correction for acceleration head h_{ai}

Calculate the head required to accelerate the fluid in the inlet piping during each pulsation cycle from

$$h_{ai} = \frac{700 \cdot N \cdot Q}{Z} \sum \frac{l}{d^2} \qquad \text{(m)}$$

where l (m) is the length of line of diameter d (mm).

12.3 Calculation of available NPSH

Calculate the available NPSH at the pump suction flange from

$$\text{Available NPSH} = \text{Basic NPSH} - h_{ai} \qquad \text{(m)}$$

12.4 Correction to NPSH for fluid properties

The following corrections are for use in discussions on layout and system design. They would not normally be part of the specification to the pump supplier.

(a) Suspended solids content

The behaviour with slurries is notoriously difficult to predict; the only reliable test is to run the pump on the plant. Non-Newtonian fluids should always be the subject of experimental tests. For suspensions of hard crystalline solids use the following empirical rules as a guide.

(i) Very fine suspensions where all the particles are less than 3 μm are best treated as viscous liquids – see Fig. 12.2.

(ii) For slurries having a range of particle sizes take

$$\text{NPSH}_{\text{slurry}} = \frac{1}{1 + x} \cdot \text{NPSH}_{\text{carrier}}$$

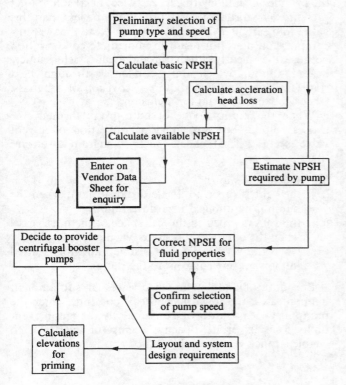

Fig. 12.1. Inlet conditions

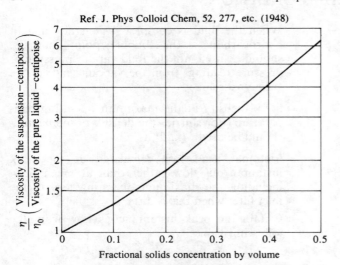

Ref. J. Phys Colloid Chem, 52, 277, etc. (1948)

Fig. 12.2.   Estimate of viscosity of fine suspensions

where

x = fractional content by volume of solid particles and the NPSH for the carrier liquid has already been corrected for viscosity.

(b)   Viscosity

An increase in viscosity increases the NPSH required. For high values of viscosity the losses across valves are approximately proportional to the liquid velocity. Consequently the NPSH required is directly proportional to viscosity at constant speed and flow.

However, it is normal practice to reduce the mean plunger speed as viscosity increases thus reducing the NPSH required, see Fig. 12.3.

(c)   Gas content

As pressure reduces across the pump inlet valves on the inlet stroke, dissolved gas may be released from solution

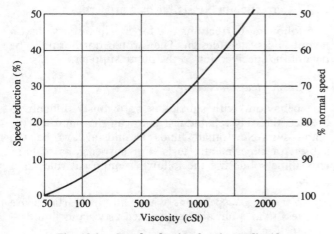

Fig. 12.3.   Speed reduction for viscous liquids

if the available NPSH is too small. Re-solution is relatively slow so that occluded gas bubbles persist. Bubble formation becomes significant when the pressure falls below 70 per cent of the pressure at which the liquid is saturated with gas. The effect is to reduce volumetric efficiency.

12.5   Estimation of NPSH required by the pump

The NPSH required by a pump is determined by its detailed design and its operating speed.

As a first estimate take

$$\text{NPSHR} = 5U^2 + \frac{0.12 P_d^{0.75}}{\rho}$$

for liquids of low viscosity (<12 cSt)

It is common practice to supply the NPSH needs of large reciprocating pumps by inlet booster pumps of the centrifugal type. This enables the reciprocating pumps to run at high speeds with resulting lower capital costs.

12.6   Priming considerations

For complete priming the pressure at the pump inlet should be sufficient to open both inlet and discharge valves against their spring forces.

The discharge pressure at start-up should be less than the pressure reached if a cylinder fills with vapour and the plunger acts as a compressor. As the clearance volume is likely to be of the same order as the swept volume, the maximum absolute pressure developed in the cylinder is unlikely to exceed twice the absolute inlet pressure. The discharge pressure should be less than this value to clear the cylinder of vapour. Note that vapour compression does not result in immediate condensation because it is nearly adiabatic and hence yields superheated vapour which only condenses through heat transfer. This condition requires provision of a bypass. Check the line diagram for inclusion.

When the pump does not auto-prime through the process piping but is primed by operation of a vent valve (see Fig. 11.5), check the line diagram to ensure that:

(a)   the vent is shown;
(b)   the vent is piped to a safe disposal point when the pump is handling a hazardous liquid;
(c)   the vent is piped to the gas space of the inlet vessel when the inlet vessel is under vacuum;
(d)   when the vent is piped, a visible indicator is fitted in the vent line adjacent to the pump.

For duties handling liquefied gases at sub-ambient temperatures, the inward heat-leak continues after the pump has been shutdown, consequently the pump eventually loses its prime. Specify lagging for inlet lines to enable a quick restart of the pump.

CHAPTER 13.   POWER RATING

The power rating sequence is modelled in Fig. 13.1.

13.1   Power required by pump

(a)   Mechanical efficiency, η_m

This is defined by

$$\eta_m = \frac{\text{Theoretical hydraulic power} \times 100}{\text{Power input to pump}} \quad (\%)$$

For preliminary estimation take

$$\eta_m = \frac{100}{1.3 + \{5/(Q \cdot P)\}} \quad (\%)$$

(b)   Overall efficiency η (per cent)

The overall pump efficiency is defined as

$$\eta = \frac{\eta_v \cdot \eta_m}{100} \quad (\%)$$

Obtain an estimate of η_v from Fig. 11.4.

(c)   Calculation of absorbed power, E

$$E = \frac{10Q \cdot P}{\eta} \quad (\text{kW})$$

Multiply the mean power requirement for the pump by the following additional factors as appropriate

Belt drive	1.07
Gearbox drive (transmitting E kW)	$1.04 + \dfrac{0.2}{E}$

13.2   Driver type

(a)   Electric motors

Where two or three pumps run continuously in parallel and cannot be provided with gas-loaded accumulators, the preferred driver type is a synchronous electric motor. The phase differences between crankshafts can then be constant and periodic pulsation beats avoided.

For an electric motor drive select the next highest power rating in kW irrespective of speed from the following sequence

22	30	37	45	55	75	90	110	132	150
185	220	250	275	300	350	400	450		

For larger motors consult a specialist or the motor manufacturer as necessary for advice on appropriate motor rating and type.

This power rating should be used only for a *preliminary* estimate of the electrical distribution load. Check

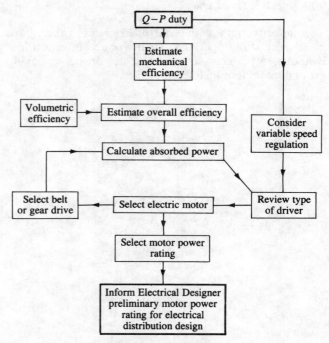

Fig. 13.1.   Power rating

the required power rating and speed after selection of pump using the manufacturers data.

Standard motors have a starting torque exceeding 150 per cent full load torque which is adequate for these reciprocating pumps.

Because the crankshaft speed is low, some manufacturers offer an integral gearbox built into the crankcase whilst others recommend a belt drive. The appropriate motor speed is the next standard lower than

$$\left(\frac{9500}{E}\right)^{2/3} \quad (\text{rev/s})$$

where E is the motor rating in kW

Belt drives rated above 220 kW require special investigation. If synchronous drive is required, specify a toothed belt.

(b)   Steam turbines

Geared steam turbines for powers less than 600 kW should be restricted to non-condensing operation because the costs of blade stressing reviews is usually not justified.

(c)   Special drivers

The following drivers require investigation outside the scope of this design guide:

(a)   geared gas turbines;

(b)   direct-coupled or geared, diesel or dual-fuel, engines;

(c)   hydraulic turbines.

(d)   Variable speed units

Torque-converters with controlled guide vanes have been successfully operated. Their average insertion loss is about 15 per cent so that their use on power saving grounds is seldom justified.

Fluid couplings or electromagnetic slip couplings are generally unsuitable because in most applications the mean torque does not vary with speed.

For variable frequency electric motors consult a specialist or the manufacturer for advice on type and range of application.

CHAPTER 14. CASING PRESSURE RATING

The process for the determination of casing pressure rating is modelled in Fig. 14.1.

14.1 Maximum discharge pressure

Calculate maximum discharge pressure, P_{dmax}, from

$$P_{\text{dmax}} = P_{\text{R}} + \frac{\rho_{\text{m}}}{10.2}(h_{\text{sd}} + h_{\text{rd}}) \qquad (\text{bar g})$$

where

P_{R} = set pressure of relief valve on delivery vessel (bar g)

ρ_{m} = maximum liquid density (kg/l)

h_{sd} = maximum static head from pump to the free liquid surface in the delivery vessel (m)

h_{rd} = maximum frictional head loss in the delivery pipe system at the peak instantaneous flow (see section 12.2) (m)

14.2 Discharge pressure relief device

A discharge pressure limiting device is required unless there is no possibility of a restriction occurring in the discharge system between the pump and a vessel already protected by a pressure relief valve. Electric motor overload protection alone is *not* sufficient.

The conventional pressure limiting device is a relief valve in the discharge pipework.

The relief valve should be sized for at least 110 per cent of the pump capacity at maximum pump speed.

14.3 Calculation of cylinder outlet losses

As a preliminary estimate take the pressure increment due to pump head outlet losses, P_{H}, as twice the pressure corresponding to the inlet NPSH required by the pump (see section 12.5). Check the outlet losses again after final selection of the pump using the manufacturers data.

14.4 Maximum achievable casing temperature

Long-term operation against a closed discharge isolation valve is usually possible when the motor can provide the power needed by the pump operating with all the flow passing through the pressure relief valve. The consequent rise in temperature is sufficiently slow to rely upon remedial action by plant operators.

If both inlet and discharge isolation valves are closed, pump operation can sharply raise the casing pressure,

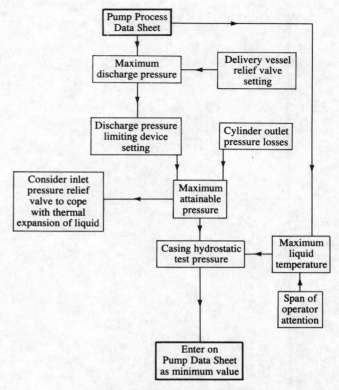

Fig. 14.1.   Casing pressure rating

through thermal expansion of the liquid. Consider specifying a pressure relief valve on the pump inlet.

14.5 Casing hydrostatic test pressure

Specify the hydrostatic test pressure to be greater than 150 per cent of the maximum attainable pressure, taken as the sum of the pressure setting of the discharge pressure limiting device and the pump cylinder outlet losses.

Pumps handling hot liquid (above 175°C) require individual consideration taking into account piping loads and reduction in material strength with increase in temperature.

Specify the rating of the inlet connection to be identical to that of the discharge connection. See Fig. 11.4 for the cut-points showing change in pipe and fitting specification or standard.

37

Positive displacement pumps: rotary

CHAPTER 15.   INTRODUCTION

Part Four covers the requirements for the specification of the pump duty and selection of rotary positive displacement pumps. This classification of pumps includes, for example, gear pumps, screw pumps, lobe pumps, sliding vane pumps, mono-type pumps, and peristaltic pumps.

The system of calculations provided will allow the reader to develop the specifications of the pump duty for enquiries to be sent out to pump vendors, using the medium of the Pump Data Sheet (Appendix IX), and provide an estimation of the characteristics and requirements of the pump(s) in order that related design work (i.e., civil, piping, electrics, instruments, etc.), can proceed in parallel.

Rotary positive displacement pumps comprise the largest proportion of pumps, next to centrifugals, used on process plants, but even so, this amounts to less than 10 per cent of all pumps used. Their great field of application is for pumping oils or other liquids having lubricating properties.

For ease of treatment, the text has been grouped into chapters dealing with the preliminary choice of pump, inlet conditions, flow/head rating sequence, driver power rating, casing pressure rating, and sealing considerations. Each chapter is supported by a flow diagram providing a model of the sub-systems of calculations. The tacit assumption underlying this arrangement is that the same full sophisticated and rigorous treatment will be applied to each and every pump installation. In practice, however, the treatment applied to the specification and selection of individual pumping installations can vary over a wide spectrum from a fairly cursory examination of a limited number of critical factors, to the full and rigorous treatment assumed here. As outlined in Part One, this will depend on the critical nature of each installation.

The flow diagram shown in Fig. 15.1 provides a model of the whole system of calculations to be carried out for rotary positive displacement pumps. Each operation is cross-referenced to the relevant chapters or sections as appropriate (numbers in right hand side of boxes). As explained in Part One, the system is broken down into three stages in order to cope with the needs of different users and/or different pump installations. The first stage covers the preliminary choice of pump type and speed; the second stage covers the more sophisticated treatment, with particular reference to methods of capacity control, fluid properties, and pressure pulsations; and the third stage covers information requirements for completion of the data sheet and for the needs of other related design work (i.e., civil, piping, electrical, instruments, etc.) which are dependent on the choice of pump type and speed.

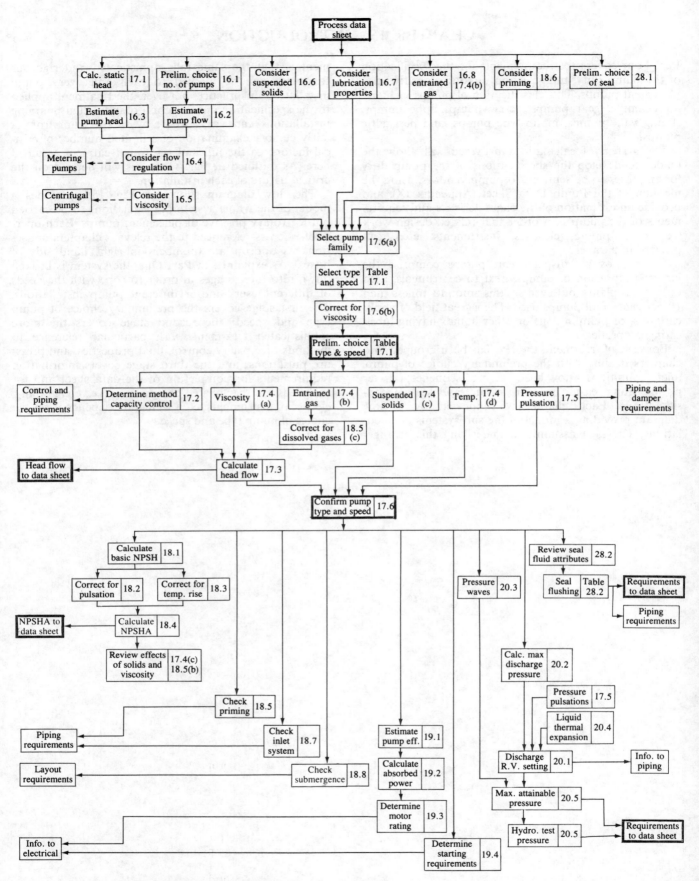

Fig. 15.1. Flowchart of the decision-making process for the selection of rotary positive displacement pumps

CHAPTER 16.   PRELIMINARY CHOICE OF PUMP

The system for the preliminary choice of rotary positive displacement pumps is modelled in Fig. 16.1.

16.1   Preliminary choice of number of pumps

Rotary pumps fall into reliability classes 4, 5, or 6 as defined in Appendix II. The common arrangement is one running pump rated at 100 per cent duty, with one identical pump as the spare either installed or available as a replacement.

16.2   Estimate of pump flow

Normal flow is taken as the largest process flow required for operation at the rated plant daily output.

Take 105 per cent of this normal flow, through one pump or divided between two pumps as considered in section 16.1, to obtain the pump flow rate Q_0 in l/sec.

16.3   Estimate of pump head

Calculate static head H_s as section 17.1. Estimate the pump head H_0 as

$$H_0 = H_s + H_r \qquad (m)$$

where H_r is the total frictional head loss in the inlet and delivery system at normal flow.

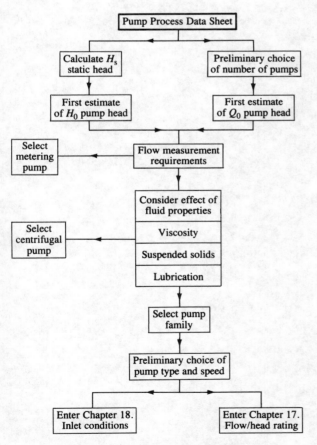

Fig. 16.1.   Preliminary choice of pump

16.4   Considerations of flow measurement

(a)   For accurate measurement of small flows, where

$$\frac{Q_0}{H_0} < 0.08 \quad \text{and} \quad Q_0 < 0.8$$

use a reciprocating metering pump – see Part Two.

(b)   For rough measurement of small flows the speed of a variable-speed gear pump gives an indication of flow when the fluid viscosity is sufficiently high to make the slip flow insignificant, taken as

$$v > 1.6 \, \frac{H_0}{Q_0}$$

16.5   Viscosity

The primary application of rotary pumps is on fluids of high viscosity. If

$$\frac{v}{Q_0^{1/2} \cdot H_0^{1/4}} < 20$$

where

v is the kinematic viscosity of liquid in centistokes

then a centrifugal pump may be a better choice – see Part Five.

16.6   Suspended solids

In order to choose the appropriate family of rotary pump construction it is convenient to classify fluids as follows.

(a)   Nominally clean liquids:
take Family 1 pumps as first choice.

(b)   Fluids carrying soft globular particles:
take Family 2 pumps as first choice.

(c)   Abrasive slurries:
Rotary pumps are inherently unsuitable for abrasive slurry duties. However, pumps of Family 3 can deal with fluids where the predominant size of the particles is less than 150 μm, although pump components will have a limited life.

16.7   Lubrication properties

If the liquid is free from particulate contaminants then the wear of internal rubbing surfaces depends on the lubrication capability of the pumped liquid.

(a)   If the liquid possesses the properties of a light mineral oil, any pump family may be selected.

(b)   For aqueous solutions or other liquids having a negligible boundary lubrication capability, select Family 2 or 3.

16.8 Entrained gas content

When gas is continuously entrained in large amounts, above 10 per cent by volume, take screwpumps from Family 2 as the first choice.

16.9 Preliminary choice of pump type and speed

With the pump duty and selected Family, enter section 17.6 to obtain pump type and speed. Usually there will be more than one possible choice.

CHAPTER 17.   SPECIFICATION OF FLOW AND HEAD RATING

The process for the specification of flow and head rating is modelled in Fig. 17.1.

17.1   Calculation of static head

All systems can be reduced to a transfer of liquid between two reservoirs or to circulation to and from one reservoir.

The steady pressure in each reservoir (P_d or P_i) can be either gas pressure on the free surface of the liquid or a point in the liquid system deliberately maintained at constant pressure by automatic control.

The static liquid head (h_s) is defined as the difference in elevation between the pump centre and either the free surface of the liquid or the constant pressure point in the liquid system.

The differential static head (H_s) across the pump is given by

$$H_s = (h_{sd} - h_{si}) + \frac{10.2(P_d - P_i)}{\rho} \qquad \text{(m)}$$

where

h_{sd} = static liquid head, from pump centre to delivery vessel free liquid surface or constant pressure point (m)

h_{si} = static liquid head, from pump centre to inlet vessel free liquid surface (m)

(Note that h_{sd} and h_{si} are negative when the appropriate liquid surface lies below the pump centre.)

P_d = gas pressure at free surface of the liquid in the delivery vessel, or the pressure at the constant pressure point in the delivery system (bar g)

P_i = gas pressure at the free surface of the liquid in the inlet vessel (bar g)

(Note that P_d and P_i are negative for vessels under vacuum.)

ρ = liquid density (kg/l)

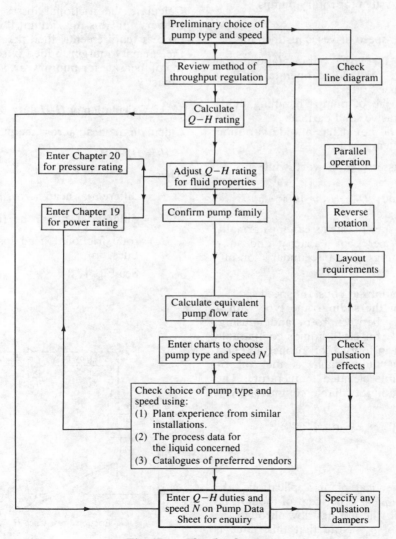

Fig. 17.1.   Flow/head rating

Check the maximum and minimum values of H_s for which the pump should still operate. Self-consistent sets of conditions are given on the Process Data Sheet.

Check the proposed layout to see if the piping system rises above the delivery vessel to form a syphon. If it does then calculate the maximum differential static head H_s with the static head h_{sd} taken to the point of highest elevation of the piping system. A syphon on the inlet piping system is ignored for this calculation because priming occurs before the pump can start.

17.2   Throughput regulation

Control is not to be construed only as automatic control; the control margin is still required if the pump is to be regulated by an operator using a manual valve. The margin can only be reduced if the range of control is reduced.

The methods of control are as follows.

(i)   Bypass control.
      This is the normal system for rotary pumps.

(ii) Speed variation.
      The cost of variable-speed drives is justified for:

   (a)   power saving for large pumps (above 200 kW) which are required to operate for long periods at reduced rates;
   (b)   extending the life of pumps handling fluids containing abrasive solid particles;
   (c)   duties where the liquid has a wide variation of viscosity or temperature.

   As a rough guide, assessment is worthwhile if

   $$\frac{Q_0 \cdot v_{\min}}{H_0} < 2 \quad \text{and} \quad Q_0 \cdot H_0 > 1000$$

   Note that variable-speed systems cannot normally operate below 20 per cent capacity and may require supplementing with a secondary 'on-off' control system.

(iii) Variable-capacity pumps at constant speed.
      Such pumps are of the sliding-vane type with a variable eccentricity between rotor and housing. They are useful only for clean liquids comparable to light mineral oil in lubrication properties. This can include heavy fluids such as fuel oil or bitumen at a suitable elevated temperature. The range of pumps readily available commercially is defined by

   $$H_0 < 80 \text{ m} \quad \text{and} \quad Q_0 < 30 \text{ l/s}$$

(a)   Bypass control

For the typical petrochemical plant application, take the normal bypass flow (q_b) as 5 per cent of the normal delivery flow (q_d), so that the normal flow through the pump is 105 per cent of q_d. Note that this value is increased by the additional margin on the pump capacity (as section 17.3) when the liquid carries suspended solids.

With this system the maximum flow through the bypass occurs at shut-off of flow to the delivery point, when the control valve is fully open.

The maximum permissible friction loss in the bypass line and the size of the control valve is fixed by the condition of delivery flow shut-off. Then

$$(H_{co} + H_{rb}) < H_s \text{ min}$$

where

H_{co} = loss in full-open control valve (m)

H_{rb} = loss in bypass line from bypass junction (m)

Both H_{co} and H_{rb} must be calculated for the flow through the bypass of q_{bmax} because q_d is then zero.

Assume $q_{bmax} = 1.1 Q_0$ and check later using the actual pump characteristic.

(b)   Variable-speed control

Calculate the frictional losses for maximum flow to delivery. If this maximum flow is not specified by process requirements, then take it as $1.05 q_d$. Note that the capacity margins added in section 17.3 are intended to compensate for pump wear, *not* to provide additional capacity.

17.3   Calculation of Q-H duty

Calculate the head across the pump from

$$H = H_s + H_r$$

where

H = differential head across pump (m)

H_s = differential static head (m)

H_r = total frictional head loss in inlet and delivery lines (m)
See Fig. 17.2

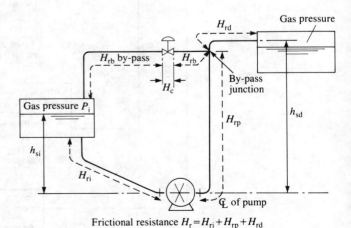

Frictional resistance $H_r = H_{ri} + H_{rp} + H_{rd}$

Fig. 17.2

Add the following cumulative margins to the basic pump flow rate Q_0:

Pump or duty classification	Margin (%)
Pumps handling nominally clean liquids	nil
Pumps handling non-abrasive slurries	10
Pumps handling abrasive slurries	(see note)
Pumps running at variable speed	nil

Pumps are normally constructed in corrosion resistant materials, consequently no design margin is added on this account.

Note. For abrasive slurries, experience from a similar installation is required. As a rough guide first assume a margin of 35 per cent (with a variable-speed drive).

17.4 Influence of fluid properties

(a) Viscosity

As viscosity increases, slip becomes small, and the capacity of the pump approaches displacement capacity. As the viscosity decreases, slip rapidly approaches the displacement capacity and the capacity of the pump drops rapidly to zero or even a negative quantity. If the pump is to operate over a wide range of viscosities, then:

(i) calculate duties for maximum and minimum viscosities and include these on the enquiry;
(ii) consider flow regulation by speed variation.

The chief problem on high viscosity fluids is filling the working chamber. This necessitates a reduction in speed with increase in viscosity – see section 17.6.b. Twin-screw pumps have the best inlet form and have a reasonable efficiency when

$$\frac{N^{5/3} \cdot v}{H \cdot Q^{2/3}} < 3000$$

provided that

$$\frac{Q}{N} > 0.46$$

Pump manufacturers generally claim an ability to handle fluid of much higher viscosity. Applications outside these limits should be supported by reference to identical satisfactory installations.

Non-Newtonian fluids such as polymer melts and dye pastes cannot be described generally (see Appendix III). For estimating efficiency and capacity use the apparent viscosity relating to the high shear leakage paths, obtained from an existing similar pump handling the identical fluid. The effect of thixotropy is important. A pump system which has been shut down with lines full may be impossible to restart unless both motor and pump are rated for very high starting torques.

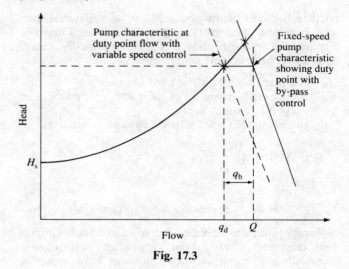

Fig. 17.3

(b) Entrained gas content

Small quantities of occluded gas simply reduce the effective mass flow capacity of the pump.

When the gas is not occluded but entrained as bubbles of large size, pumps in Family 2 or 3 are essential.

At gas contents exceeding 15 per cent by volume the bubbles collapse rapidly in the pump discharge zone, generating noise and vibration. The preferred pump type is then a twin-screw. This has multiple seals in series, enabling compression to take place over the length of the screw provided that the slip through the rotor/casing clearance is sufficiently high. A cooled external recirculation of the pumped liquid at a rate of at least

$$\frac{Q^2}{N} \text{ l/sec}$$

is required when a pump can operate at all gas/liquid ratios up to 100 per cent gas.

(c) Suspended solids

Abrasive particles are crystals of material whose intrinsic hardness is greater than the hardness of the materials used in pump construction and whose size is comparable to the clearance between nominally contacting surfaces. See Appendix IV. Freshly formed crystals possess sharp edges; those that have suffered attrition are rounded and consequently not so damaging.

A typical contaminant source is airborne dust containing sand or milled rock product. At ground level, where pumps and tanks are commonly located, the predominant particle size lies in the 70 to 140 μm band. At higher levels, at least 2 m above ground in unobstructed areas, airborne dust size lies in the 0.1 to 5 μm band. Caution is necessary when judging the abrasive nature of particles. There is a marked change in the 'feel' of hard sharp grains rubbed between the fingers when the particle size falls below about 70 μm; the material then seems smooth, not gritty. Rotary pumps in Family 1 require the pumped liquid to be nominally clean, defined by the following criteria:

  (i)   the suspended solids are not abrasive;
 (ii)   99 per cent by weight of the particles are less than $30(Q/N)^{1/3}$ μm in size;
(iii)   the suspended solid particle content is small, viz

$$X < \frac{100}{Q^{2/3} \cdot N^{1/3}}$$

where

X = solids content by volume (μm^3/m^3)

Rotary pumps in Families 2 and 3 can handle slurries containing non-abrasive particles. Such particles are essentially soft material and of globular form, typically an amorphous organic material free from contamination by rust, sand, or similar hard crystalline substance.

(d)   Temperature

Rotary pumps require close running clearances for rotor/rotor or rotor/stator seals but these have to take account of transient differential temperatures. Ensure that the operating temperature range is given in the enquiry.

If the pump is to operate over a wide temperature range the variation in slip may be abnormally wide. The preferred method of regulation is then by variable speed.

17.5   Pressure pulsation

To avoid pulsation enhancement, the length of piping between the pump ports and a large discontinuity such as a vessel should be less than 4 per cent of the fundamental wavelength (Λ) where

$$\Lambda = \frac{C}{N \cdot n} \text{ (m)}$$

C = velocity of sound in fluid (m/s)

n = number of flow pulses during one revolution of pump rotors

Calculate the level of pressure pulsation at inlet and discharge of pump from

$$\delta P = K_1 \cdot Q \cdot N \cdot \rho\left(\frac{L}{d^2}\right)$$

where

δP = pressure pulsation (peak to peak) (bar)

d = diameter of line bore (mm)

L = length of line to large discontinuity (m)

K_1 = 120 for lobe pump
       20 for mono type pump
       40 for spur gear pump (more than 8 teeth)
      200 for vane pump
      200 for peristaltic pump – planar cam type
      600 for peristaltic pump – multiple roller type
        0 for screw pump or helical gear pump

Alternatively, if the system presents no discontinuity in the piping close to the pump, then calculate the basic pressure pulse at inlet and discharge of pump from

$$\delta P = \frac{K_2 \cdot \rho \cdot C \cdot Q}{D^2}$$

where

δP = basic pressure pulse (bar)

K_2 =   2 for lobe pumps with 3 lobes
          5 for lobe pumps with 2 lobes
          5 for mono type pump
          1 for spur gear pump (more than 8 teeth)
          5 for vane pump
         20 for peristaltic pump – planar cam type
         10 for peristaltic pump – multiple roller type
          0 for screw pump or helical gear pump

For lines of varying diameter, take D as the minimum line diameter excluding short lengths ($<0.01\Lambda$).
If

$$0.04P^{0.7} < \delta P$$

then the pressure pulse is unacceptable and piping vibration is likely. Reduce the level by increasing the line diameter, or specify a pulsation damper.

17.6   Guide to pump type and speed

(a)   Choice of pump family

(i)   Family 1 (*See Table 17.1 and Fig. 17.5*). These pumps are characterized by having internal bearings, and by torque transmission between rotors by internal timing gears or by engagement between the rotors themselves.

Such pumps cannot handle liquids containing abrasive particles without suffering wear. They should be the first choice when the system is closed and provided with adequate filters to remove the particles; as exemplified by lube oil systems and by hydraulic power transmission systems.

These pumps are also suitable for nominally clean liquids when the minimum film thickness between contacting surfaces is large enough to permit operation without a filter, typically when the liquid viscosity exceeds 200 cSt and the pump head is less than 200 m.

Thin film lubrication properties become dominant when

$$\frac{N^{2/3} \cdot Q^{1/3} \cdot v}{H} < 0.7$$

and the materials of construction should then be chosen for their anti-gall or self-lubrication qualities.

Table 17.1. Family 1 (See Fig. 17.4)

Zone Code	Pump type	Maximum rotor speed N* (rev/s)	Zone Code	Pump type	Maximum rotor speed N* (rev/s)
A	3-screw	50	G1	External helical gear	25
A1	3-screw	50		Internal gear	25
	Internal gear	12		3-screw	50
A2	3-screw	50	G2	External helical gear	25
	Internal gear	12		Internal gear	25
	Vane	8		3-screw	50
			G3	External helical gear	25
B	3-screw	25		Internal gear	16
B1	3-screw	25		3-screw	50
	Internal gear	12		Sliding Vane	8
			G4	External helical gear	25
C	3-screw	16		Internal gear	16
				3-screw	50
D	3-screw	12		Sliding Vane	12
			G5	External helical gear	25
E	2-screw (internal bearings)	16		Internal gear	25
				3-screw	50
F	2-screw (internal bearings)	12		Sliding Vane	12
G	External helical gear	25			
	3-screw	50			

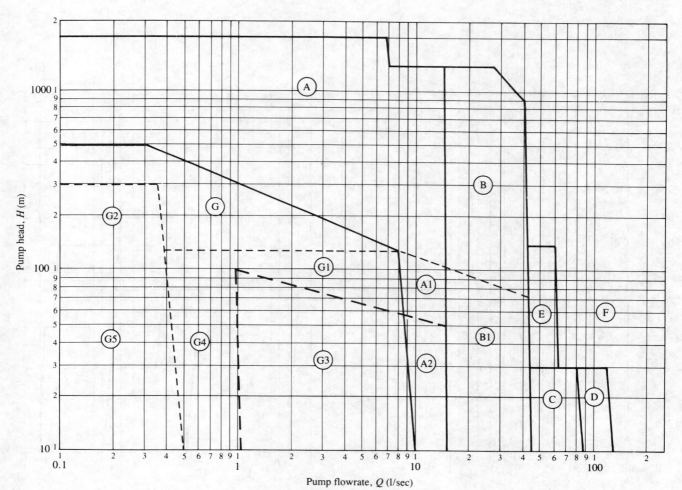

Fig. 17.4. Family 1

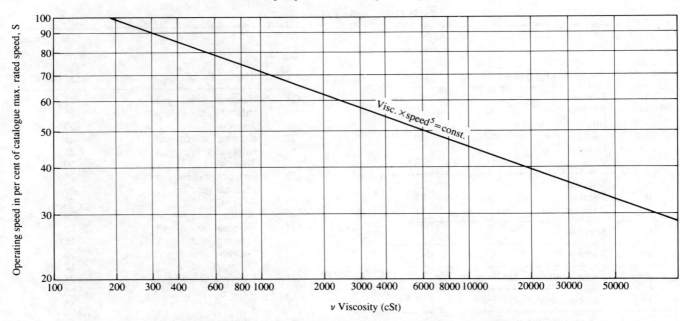

Fig. 17.5.   Typical speed reduction requirements for pumping viscous liquids for rotary pumps

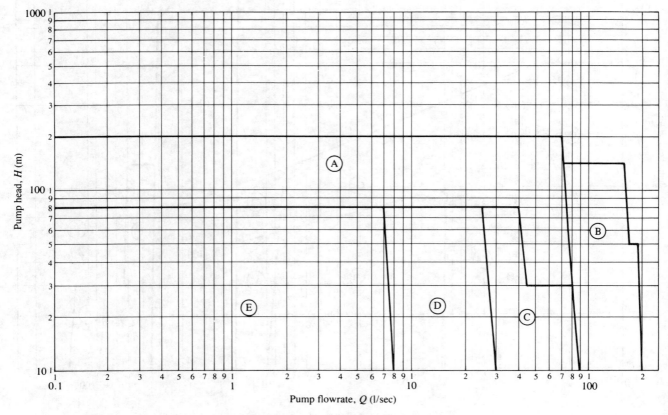

Fig. 17.6.   Family 2

(ii) Family 2 (See Table 17.2 and Fig. 17.6). These pumps are characterized by having external bearings and timing gears so that there is no contact between rotors nor any load-carrying area between rotor and casing.

These pumps can handle liquids containing small amounts of globular particles, provided that the intrinsic hardness of the particle is less than the hardness of the pump construction materials, and provided that the predominant size of the particles is not greater than 30 per cent of the clearances. See Appendix IV.

Pumps in this family are insensitive to the lubricating properties of the liquid and to the gas content of the fluid.

(iii) Family 3 (See Table 17.3 and Fig. 17.8). These pumps are characterized by having elastomeric elements which permit elastic indentation by small particles.

They can handle non-spherical particles although needle-shaped crystals will suffer attrition. On abrasive slurry duties, pumps will have a limited life. Accordingly they are constructed so that components can be readily replaced. Typically a mono type pump on rock slurry duty has a life expectancy of 400 to 1200 hours.

These pumps are satisfactory when operating with aqueous solutions.

(b) Calculation of equivalent flowrate

The charts in this guide apply for fluids of viscosity 100 cSt.

Enter the chart for the appropriate pump family with Q and H, to select pump type and maximum rotor speed N^*.

For a fluid of viscosity v cSt, determine the operating speed N from Fig. 17.5.

Note that considerations of Nett Positive Static Head (NPSH) may impose a limit on N.

Table 17.2. Family 2 (see Fig. 17.6)

Zone Code	Pump type	Maximum rotor speed N^* (rev/s)
A	2-screw (external bearings)	25
B	2-screw (external bearings)	16
C	2-screw (external bearings)	25
	Lobe	8
D	Lobe	12
	2-screw (external bearings)	25
E	Lobe	16
	2-screw (internal bearings)	25

Table 17.3. Family 3 (see Fig. 17.8)

Zone Code	Pump type	Maximum rotor speed N^* (rev/s)
A	Mono type	16
B	Mono type	12
C	Mono type	8
A1	Peristaltic	1
	Mono type	16
A2	Peristaltic	0.5
	Mono type	16
A3	Peristaltic	0.25
	Mono type	16
A4	Elastomeric Vane	16
	Peristaltic	1
	Mono type	16

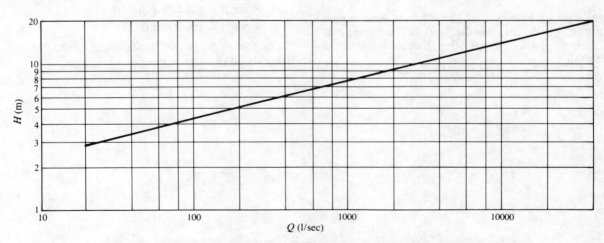

Fig. 17.7. Family 2 – Archimedean screw pumps (Supplement to Fig. 17.6).

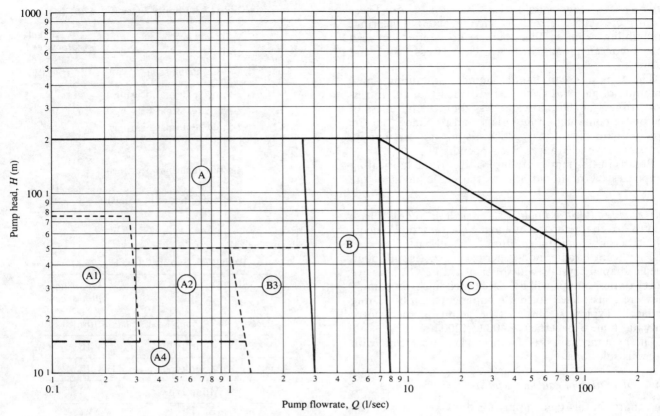

Fig. 17.8. Family 3

Then calculate the equivalent pump flowrate, Q_e, at the speed appropriate for a viscosity of 100 cSt, as

$$Q_e = \frac{N^*}{N} \cdot Q$$

Re-enter the chart to confirm or adjust N^*.

Archimedean screw pumps (see Fig. 17.7)

A special case of Family 2 is the Archimedean single screw pump which can handle heavily contaminated liquids such as sewage. Its effective limit of application is $H < 1.4Q^{1/4}$.

These pumps are physically large. As a guide to their dimensions take

$$L = 2.5H^{1.3} + 0.04Q^{0.4}$$

$$\frac{L^2}{80} < D < \frac{Q^{1/2}}{14}$$

$$N = \frac{0.7}{D}$$

where

  L = rotor wetted length (m)
  D = rotor diameter (m)

Take the angle of the shaft centre to the horizontal as roughly 30 degrees.

CHAPTER 18. INLET CONDITIONS

The process of confirming that the inlet conditions are suitable for the preliminary choice of pump is modelled in Fig. 18.1.

18.1 Calculation of basic NPSH

Basic NPSH

$$= h_{si} - h_{ri} + \frac{10.2}{\rho}\left(\frac{B}{1000} + P_i - P_v\right) \quad (m)$$

where

h_{si} = static liquid head over pump inlet (m)

This is measured from the lowest liquid level to the pump centre and is negative when the liquid level is below the pump centre. See Fig. 18.2. The lowest level is not necessarily the extreme value but is the one at which the pump should still operate.

B = Minimum barometric pressure at pump location (mbar)

Use 0.94 of the mean barometric pressure from local meteorological data to allow for extreme weather conditions.

P_i = Minimum working value of the gas pressure on the free liquid surface in the inlet vessel. This is negative for vessels under vacuum (bar g)

P_v = Vapour pressure of the liquid at the maximum operating temperature (bar a)

When the fluid is complex – generally a light hydrocarbon – the 'bubble point' pressure should be used, not the Reid vapour pressure nor pressure readings from the vapour space of the supply vessel.

h_{ri} = Frictional head loss in the inlet piping system (including the maximum operating loss in the inlet filter)

ρ = Liquid density at the maximum operating temperature (kg/l)

Note

(1) For liquids at boiling point the function in the brackets is always zero so that the basic NPSH becomes simply

$$h_{si} - h_{ri} \quad (m)$$

(2) When there is heat transfer the liquid is not in an equilibrium state.

(a) For an internal heat source, e.g., a calandria in an evaporator, the column of liquid/vapour in the calandria tubes reduces the pressure at the calandria base and induces circulation. For pumps the effective quiescent liquid surface is then very nearly the midpoint of the calandria – see Fig. 18.2.

(b) For an external heat source, e.g., nominally refrigerated liquid entering a storage vessel, small differences in temperature occur between the liquid and its vapour. Check that such liquid entry connections are above the maximum liquid level to permit 'flashing' and thus obtain temperature equilibrium.

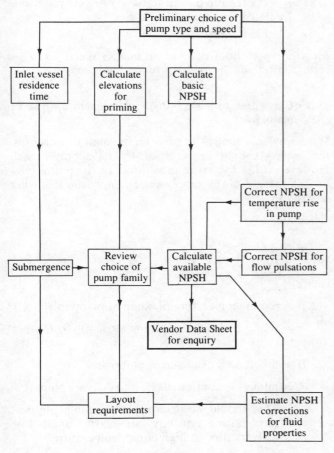

Fig. 18.1. Inlet conditions

18.2 Correction to basic NPSH for flow pulsation

The pulsations in flow delivered by some types of rotary pump cause similar variations in fluid velocity in both inlet and discharge piping. Consequently the basic NPSH calculated in section 18.1 should be corrected for:

(a) frictional head loss in the inlet pipe system at peak instantaneous flow rather than maximum mean flow;

(b) head required to accelerate the fluid in the inlet piping during each pulsation cycle.

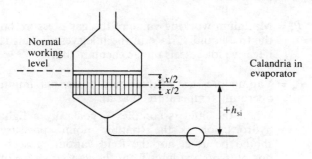

Normal working level

Calandria in evaporator

$+h_{si}$

$x/2$
$x/2$

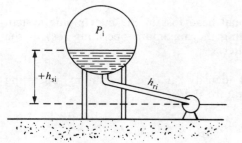

P_i

$+h_{si}$

h_{ri}

Inlet vessel under gas pressure

Note that for vacuum vessels P_i is reckoned negative

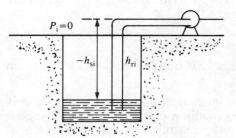

$P_i = 0$

$-h_{si}$ h_{ri}

Atmospheric sump

Note that h_{ri} is reckoned negative

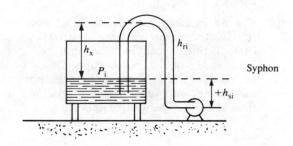

h_x h_{ri}

P_i

$+h_{si}$

Syphon

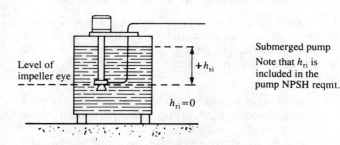

Level of impeller eye

$+h_{si}$

$h_{ri} = 0$

Submerged pump

Note that h_{ri} is included in the pump NPSH reqmt.

Fig. 18.2.   Examples of alternative suction arrangements

(a)   Frictional head correction

Correct the friction head loss in the inlet piping H_{ri} for the peak flow obtained from the following table

Pump type	$\dfrac{Peak\ flow}{Mean\ flow}$
Any type with damper on inlet	
Screw pumps	1.0
Gear pumps with helical gears	
Mono type pumps	1.03
Vane pumps	1.05
Lobe pumps – 2 lobes	1.25
– 3 lobes	1.15
Peristaltic pumps – planar cam type	2.0
– multiple roller type	1.5

(b)   Acceleration head

Calculate the acceleration head loss in inlet piping from

$$H_{ai} = \frac{\delta P}{\rho} \quad (m)$$

where

δP = inlet pressure pulsation/basic pressure pulse calculated in section 17.5

ρ = liquid density

for a pump without damper on inlet. Take H_{ai} as zero for a pump that is provided with an inlet damper.

18.3   Correction to basic NPSH for temperature rise at pump inlet

Where bypass control is used for regulation purposes, the bypass should be returned to the supply vessel. However, if the bypass returns direct to the pump inlet branch instead of the supply vessel, apply the following correction.

Calculate

$$\Delta T_i = \frac{H}{102 \cdot \xi} \cdot \frac{100}{\eta} \cdot \frac{q_b}{q_d}$$

where

ΔT_i = rise in temperature of liquid at pump inlet (°C)

η = pump hydraulic efficiency at point (Q, H) (per cent)

H = differential head across pump (m)

q_d = minimum continuous flow to delivery point (l/s)

This should be determined by some physical mechanism, typically an alarm upon low delivery flow or high liquid temperature.

q_b = corresponding flow through bypass (l/s)

Q = corresponding flow through the pump (l/s)

ξ = liquid specific heat (kJ/kg °C)

Then find the vapour pressure corresponding to the specified inlet temperature plus the increment ΔT_i.

Recalculate the basic NPSH using this value of vapour pressure together with the head loss in the inlet line system recalculated for flow q_d, not the maximum flow through the pump. If this recalculated value of basic NPSH exceeds the original value then ignore the correction.

Note that when there is no flow to the delivery point the temperature of the inlet liquid steadily increases until balanced by radiation and convection losses.

18.4 Calculation of available NPSH

Calculate the available NPSH by subtracting the corrections given in sections 18.2 and 18.3 from the basic NPSH.

Although the NPSH value so obtained corresponds to *maximum* flow through the pumps it is the proper value to specify for the rated duty on enquiries.

18.5 Effect of fluid properties on NPSH

(a) Pump characteristic

For rotary pumps the 'slip' returns hotter fluid to the inlet, thus raising the effective inlet fluid temperature and vapour pressure. For a given pump

$$\Delta T_i \propto \frac{H^2}{\xi \cdot \eta \cdot N \cdot v}$$

(b) Viscosity

An increase in viscosity increases the NPSH required. However, this effect is generally countered by a reduction in pump rotational speed. Typically, for a viscosity change from 100 to 10000 cSt the NPSH required will increase by a factor of two when speed is reduced in accordance with section 17.6(b).

(c) Dissolved gas

Where the inlet vessel is a gas scrubber, the liquid is likely to be saturated with the gas. For such duties the calculated basic NPSH may be deceptively high; the pump performance being affected by gas release rather than by the vapour release upon cavitation. The audible and destructive effects of cavitation are mitigated by the presence of the gas but a correction to the basic NPSH is required to avoid a shortfall in pump performance.

Apply the following procedure.

(i) Use gas solubility data to obtain the relation between absolute pressure and the gas release in actual m³/m³ liquid. Gas release is normally insensitive to temperature: exceptionally some systems may need evaluation both at maximum and minimum pumping temperatures.

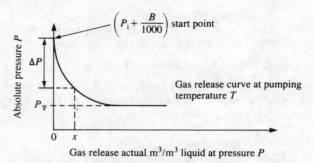

Fig. 18.3. Gas release/pressure relation

Note that boiling liquids can be assumed to contain no dissolved gas.

(ii) Take the effect on pump performance as a fall of 5 per cent in liquid capacity, corresponding to a gas content at inlet of 5 per cent by volume. From this gas release/pressure relation obtain the pressure reduction (ΔP) at which the actual gas to release, measured at pressure P, is x per cent by volume (see Fig. 18.3). Because only a small proportion of the liquid actually passes through the low pressure zone within the pump, x is dependent on the pump type and construction but can be taken as 0.5 for a representative value.

(iii) Recalculate the basic NPSH, replacing

$$\left(P_i + \frac{B}{1000} - P_v \right) \quad \text{by} \quad \Delta P$$

Note that this step assumes that the gas content gives a pseudo-vapour pressure of

$$\left(P_i + \frac{B}{1000} - \Delta P \right)$$

18.6 Priming

(a) The static liquid level at pump inlet (H_{si}) is the height to which the liquid will rise above pump centre

$$H_{si} = h_{si} + \left(\frac{10.2 P_i}{\rho} \right) - h_{ri} \quad \text{(m)}$$

where

h_{ri} = frictional loss in inlet system at normal flow. Note that this term is included because general operational policy requires that the standby pump can start whilst the operating pump continues to run (m)

h_{si} = the difference in elevation between the pump centre and the lowest working level in the inlet supply vessel (m)

Note that h_{si} is negative when the liquid level is below the pump centre.

P_i = minimum working gas pressure in the inlet vessel (bar g)

Note that P_i is negative for vessels under vacuum, or taken as zero if the priming vent is piped to the gas space of the inlet vessel.

ρ = maximum liquid density (kg/l)

(b) If the static liquid level in the inlet vessel is unavoidably below the highest point of the pump casing, then use:

(i) *A self-priming pump*
True self-priming action is inherently available only in peristaltic and archimedean screw pumps.
Elastomeric elements which rub continuously cannot be run dry but can run on 'snore' where sufficient liquid for lubrication is retained by the construction of the pump body. Ensure that the line diagram shows a feed for initial start-up and to replace evaporation losses.

(ii) *An external priming arrangement*
Once primed, all rotary pumps will work with a suction lift; however, upon inadvertent stop they will deprime. Such pump arrangements cannot be used for autostart duties.

(iii) *A vertical immersed pump.*

(c) When the pump does not auto-prime through the process piping but is primed or cooled down by operation of a vent valve, check the line diagram to ensure that:

(i) the vent is shown;
(ii) the vent is piped to a safe disposal point when the pump is handling a hazardous liquid;
(iii) when the inlet vessel is working under vacuum the vent is piped to the gas space of the inlet vessel;
(iv) when the vent is piped then a visible indicator is fitted in the vent line adjacent to the pump.

18.7 Inlet system layout

The inlet pipe bore should never be less than the bore of the pump casing connection. The recommended maximum inlet line liquid velocity is

$$\frac{2.8}{v^{1/5}} \quad \text{(m/s)}$$

where v is the liquid viscosity in cSt at the lowest normal operating temperature.

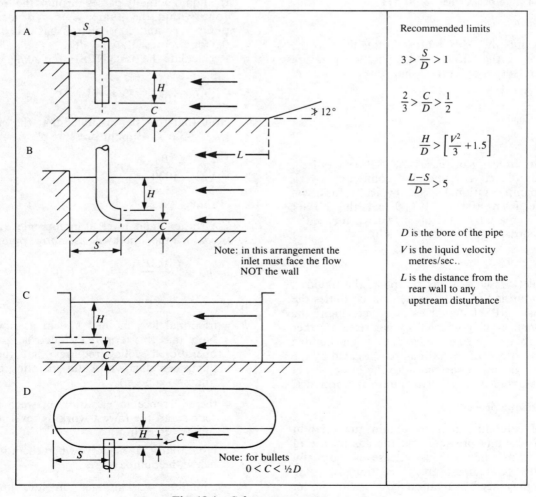

Recommended limits

$$3 > \frac{S}{D} > 1$$

$$\frac{2}{3} > \frac{C}{D} > \frac{1}{2}$$

$$\frac{H}{D} > \left[\frac{V^2}{3} + 1.5\right]$$

$$\frac{L-S}{D} > 5$$

D is the bore of the pipe

V is the liquid velocity metres/sec..

L is the distance from the rear wall to any upstream disturbance

Note: in this arrangement the inlet must face the flow NOT the wall

Note: for bullets
$0 < C < \frac{1}{2}D$

Fig. 18.4.   Submergence arrangements

18.8 Submergence

Submergence is the depth of the inlet opening beneath the free liquid surface. If the submergence is too small there is a risk of vortex formation and gas entrainment. Check the proposed layout against the approved arrangements shown in Fig. 18.4 using the liquid velocity corresponding to *maximum* flow.

(*a*) The effect of liquid viscosity is pronounced but not reliably predictable. As a rough estimate take

$$V_c = 0.15(v^{1/3} - 1) + V_w$$

where

V_w = critical inlet velocity for a given submergence in cold water (m/sec)

v = kinematic viscosity of liquid at operating temperature (when it exceeds unity) (cSt)

V_c = critical inlet velocity for the same submergence in the liquid (m/sec)

(*b*) The arrangements shown in Fig. 18.4 assume that the vessel or sump is large. For pumps drawing from small tanks, the submergence should be sufficient to suppress the effects of the fresh liquid which enters the tank to maintain the liquid level.

A useful parameter is the residence time λ, defined as the time taken in seconds at constant maximum flow for the liquid level to fall from the effective undisturbed level to the topmost point of the vessel exit branch.

Then:

(i) The residence time λ should not be less than 100 seconds without satisfactory operational experience from an identical system.

(ii) When $100 < \lambda < 200$ increase the calculated minimum submergence by the factor

$$\left(\frac{200}{\lambda}\right)^2$$

(iii) When $\lambda > 200$, the effect of the vessel size is neglected.

(*c*) Check the method of introducing the supply liquid into the tank.

(i) For $\lambda < 200$, the supply is best admitted over a weir having a suppressed nappe.

(ii) For $\lambda > 200$ a simple pipe connection is satisfactory provided that the maximum inlet velocity is 0.6 m/s and arrangements embodying deliberate swirl are avoided.

(*d*) Check the proposed hydraulic design.

(i) If swirl is deliberately induced, then compare the proposed layout with a geometrically similar layout already known to work satisfactorily. The geometric scale critical submergence will then be nearly correct for equal velocities.

(ii) Vessels with tangential inlet or central bottom exit invariably require a vortex suppressor. Vortex suppressors are difficult to design except by model tests; simple crosses in exit lines from vessels are ineffective.

CHAPTER 19.   POWER RATING

The power rating sequence for the pump driver is modelled in Fig. 19.1.

19.1   Estimation of pump efficiency

It is important to distinguish the following definitions of efficiency.

(a) 'Pump efficiency' includes the effects of liquid viscosity, solids content, the frictional losses in the gland, capacity loss in gland flushing arrangements where the flushing liquid is taken from some point in the pump, and bearing losses.

   This definition yields the power requirement at the shaft coupling.

   Note that losses in the driver and the transmission between pump and its driver (e.g., belt) transmission are not included in the efficiencies defined for the pump.

(b) 'Overall efficiency' is the combination of the 'pump efficiency' and the 'driver efficiency'.

   This definition yields the power requirements at the input terminals of the electric motor and is useful where the motor is integral with the pump.

(c) 'Hydraulic efficiency' refers only to the intrinsic performance of the rotor/stator combination.

The hydraulic efficiency is estimated as follows.

(a) For gear and screw pumps on high viscosity liquids where the effective Reynolds number is low and operation is in the laminar flow regime where

$$\frac{H}{v^2} \cdot \left(\frac{Q}{N}\right)^{2/3} < 0.1$$

estimate the hydraulic efficiency using Figs 19.2 and 19.3, respectively.

(b) For lobe pumps on low viscosity liquids, estimate the hydraulic efficiency using Fig. 19.4.

19.2   Calculation of absorbed power

$$E = \frac{10 \cdot Q \cdot P}{\eta}$$

alternatively

$$E = \frac{0.981 M \cdot H}{\eta}$$

where

E = absorbed power (kW)

M = mass flow through pump (kg/s)

Q = volume flow through pump (l/s)

H = differential head across pump (m)

P = differential pressure across pump (bar)

η = pump efficiency (per cent)

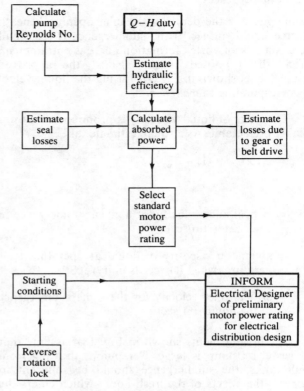

Fig. 19.1.   Driver power rating

19.3   Determination of driver power rating

Multiply the maximum power requirement of the pump by the following factors.

(a) Pump manufacturing and test measurement tolerances

   Efficiency $1.02 + \dfrac{2}{\eta}$

(b) Transmission factors

Belt drive	1.07
Gearbox drive (transmitting E kW)	$1.04 + \dfrac{0.2}{E}$
Variable speed slip coupling (transmitting E kW)	$1.02 + \dfrac{0.06}{E^{1/2}}$

(c) Cyclic power demand

Peristaltic pump – planar cam	1.3
multiple roller	1.1

(d) Emergency service

   Certain duties should be performed without risk of trip due to motor overload. Ensure that the driver power rating is adequate for continued operation against the discharge pressure relief valve, viz., against at least 110 per cent of the pressure setting of this valve.

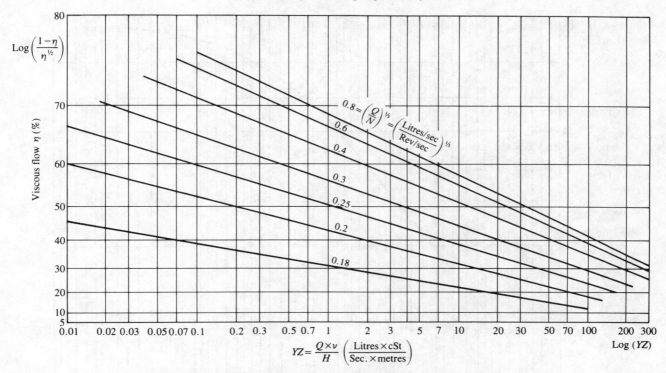

Fig. 19.2. **Generalised efficiency curves for gear-type pumps**

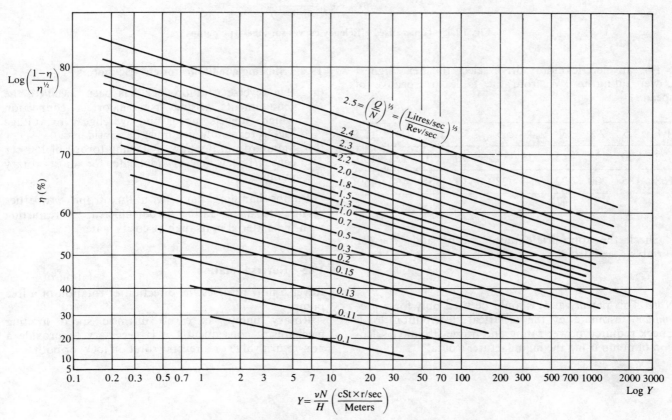

Fig. 19.3. **Generalised efficiency curves for screw pumps**

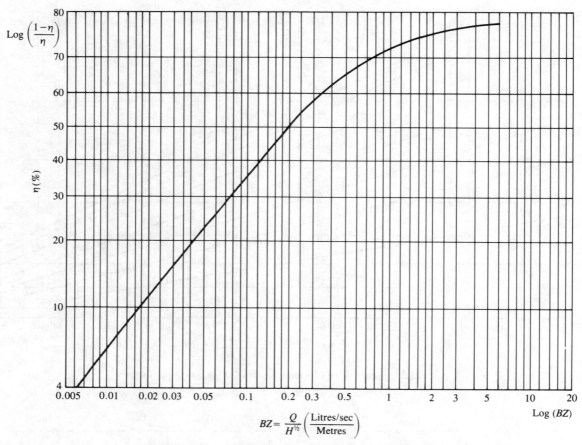

$$BZ = \frac{Q}{H^{1/2}} \left(\frac{\text{Litres/sec}}{\text{Metres}} \right)$$

Fig. 19.4.   Generalised efficiency curves for lobe-type pumps

For an electric motor drive select the next highest power rating in kW from the table, irrespective of speed

1.1	1.5	2.2	3	4	5.5	7.5	11	15	18.5
22	30	37	45	55	75	90	110	132	150
168	185	200	222	250	280	315	355	400	450

Check the appropriate motor rating and type for variable speed drives.

Note. This power rating should be used only for a *preliminary* estimate of the electrical distribution load. Check required power rating and frame size after selection of pump using the manufacturer's data.

19.4   Starting conditions for electric motors

(a)   If a reverse rotation lock has been specified, the requirements arising from incorrect connection which subjects motor and switchgear to voltage lock rotor conditions must be considered.

(b)   Standard motors have a starting torque of 150 per cent FLT which is adequate for most rotary pumps.

Pumps handling fluids with thixotropic properties require the driver rating to be sufficient for restarting with the material in its high viscosity state.

19.5   Reverse rotation

Reverse flow reverses the direction of rotation of a free pump.

Rotary pumps have no intrinsic speed limiting mechanism; consequently it is necessary to provide a non-return valve or a reverse rotation lock, or both.

CHAPTER 20. CASING PRESSURE RATING

The process for the determination of casing pressure rating is modelled in Fig. 20.1.

20.1 Discharge pressure relief valve rating

A pressure relief valve or bursting disc shall be installed at the pump discharge except for the following pump types:

(a) Archimedean single-screw pump;
(b) vane pump with elastomeric vanes;
(c) peristaltic pump whose elastomer element is totally enclosed;
(d) any pump whose driver incorporates a reliable torque limiter;
(e) any system where there is no valve or restrictor in the inlet and discharge lines between pump and vessels and *no* possibility of blockage.

The relief valve should be sized for at least 120 per cent of the full pump displacement capacity at maximum pump speed. This 20 per cent margin allows for the:

(a) tolerance permitted on performance;
(b) variations in nominally identical replacement components owing to manufacturing tolerances.

The valve set pressure should be greater than 110 per cent of the sum of the maximum discharge pressure and the pulsation pressure from section 17.5.

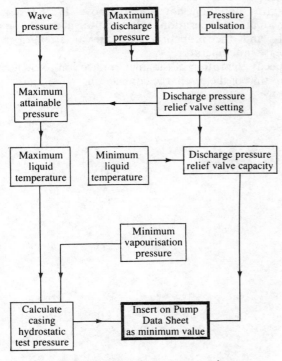

Fig. 20.1. Casting pressure rating

20.2 Calculation of maximum discharge pressure P_d^*

$$P_d^* = \frac{\rho_m}{10.2}(H_s + H_{rp}) + P_r \qquad \text{(bar g)}$$

where

ρ_m = maximum liquid density (kg/l)

h_s = maximum static head due to the elevation of the free surface of the liquid in the delivery vessel (m)

H_{rp} = frictional head loss in the delivery lines for the maximum flow rate, corrected for the peak instantaneous flow (m)

P_r = set pressure of relief valve on the delivery vessel (bar g)

20.3 Pressure waves

Pressure waves arise from changes in flow in long pipelines.

Fast changes in flow are defined by

$$t < \frac{3L}{C}$$

where

L = the total length of the delivery pipework (m)

C = the velocity of sound in the liquid at operating temperature and pressure (m/sec)

t = the time for the valve controlling the flow to travel from full open to shut (sec)

Such fast changes usually arise from:

(a) emergency flow shut-off by a solenoid trip valve;
(b) manually-operated cocks or similar quick-acting valves.

A rotary pump has substantially positive displacement action. When the driver is a direct-on-line (DOL) start electric motor the run-up time can be short, of the order of 2 seconds; consequently both start and trip of the pump may be regarded as a fast change.

For such fast changes of flow in pipelines, the head rise is given by

$$\Delta h = K \cdot v \cdot C$$

where

Δh = incremental head rise due to wave (m)

v = velocity of liquid in delivery pipeline at maximum flow (m/sec)

If the piping varies in diameter take the equivalent velocity as

$$\frac{\sum (v \cdot l)}{L}$$

where v is the liquid velocity within a line-section of length l of the total line length L.

C = velocity of sound in liquid at operating temperature and pressure (m/sec)

Note that entrained gas bubbles markedly decrease the effective velocity of sound in the liquid.

K = a factor taken as 0.09 for typical steel pipes.

For other pipeline materials take K as

$$\frac{0.1}{\left(1 + \dfrac{m \cdot \rho \cdot C^2}{1000 \cdot E}\right)^{1/2}}$$

where

$$m = \frac{\text{Internal pipe diameter}}{\text{Radial thickness of pipe wall}}$$

E = Young's modulus (MN/m^2)

20.4 Pressure due to liquid thermal expansion and vaporization

If a pump handling a liquefied gas at sub-ambient temperatures is completely isolated whilst full of liquid, the heat-leak from atmosphere can raise the pressure in the pump.

The pressure rise is not significant when:

(a) the pump is fitted with a soft-packed gland;
(b) a gas-loaded vessel acting as an accumulator or a pulsation damper is provided within the system isolated by the block valves.

Otherwise check the line diagram to ensure that a relief valve is fitted on the pump side of the inlet isolation valve. Such a relief valve should be set at the lower of the following values:

(a) 67 per cent of the pump casing hydrostatic test pressure;

(b) 90 per cent of the static pressure rating of a single mechanical seal.

The release of a small quantity of liquid is sufficient to relieve the pressure rise due to liquid thermal expansion. A continuing heat-leak may then begin vaporization of the liquid. If continuous fluid leakage is undesirable then the calculated maximum attainable pressure of both casing and seal shall exceed the equilibrium saturation pressure at the expected ambient temperature.

An allied phenomenon is the rise in liquid temperature upon operation of a pump with both inlet and discharge valves shut. When rotary pumps have a pressure relief valve connected across the pump within these block valves there is no immediate pressure excursion. The process effect of the pump running without forward flow is normally immediately noticeable and the rate of pressure rise is slow so that the pump can be manually tripped. An automatic trip is not required unless the pump is intended for autostart or remote start, and there is no associated process warning for low flow or high temperature.

Where the tank is isolated by an emergency trip valve in the pump inlet line, check the line diagram to ensure that the pump driver is simultaneously tripped.

20.5 Casing hydrostatic test pressure and maximum attainable pressure

Specify the hydrostatic test pressure for the casing or casing discharge section as 150 per cent of the maximum attainable pressure, defined as the sum of the pressure setting of the discharge relief valve and the calculated wave pressure.

Pumps handling hot liquid (above 175°C) require individual consideration taking into account piping loads and the reduction in material strength with increase in temperature.

Specify the rating of the inlet connection to be identical to that of the discharge connection.

Centrifugal pumps

CHAPTER 21. INTRODUCTION

Part Five covers the requirements for general and special purpose centrifugal pumps.

The system of calculations provided will allow the reader to develop the specifications of the pump duty for enquiries to be sent out to pump vendors, using the medium of the Pump Data Sheet (Appendix IX), and provide an estimation of the characteristics and requirements of the pump(s) in order that related design work (i.e., civil, piping, electrics, instruments, etc.) can proceed in parallel.

Centrifugal pumps account for a very large proportion (upwards of 80 per cent) of all pumps used on process plants and this clearly places Part Five in a special category of its own. Many engineers will find that their experience of pumps other than centrifugal is so small as to be almost non-existent. The proliferation of centrifugal pumps is such that one major process company has developed a computer program to carry out the system of calculations for the selection and specification of the pumps detailed here on an almost routine basis.

Further, in the case of centrifugal pumps, the interaction of the pump on the process system, and vice versa, is much more critical than is the case with positive displacement pumps. A full and thorough understanding of this interaction between pump and process system is an essential requirement for the successful plant designer. To this end an Addendum has been included, comprising Section Three of the EEUA Handbook No. 30: *A Guide to the Selection of Rotodynamic Pumps*, entitled 'Pump and system combined'. The less experienced reader is strongly advised to read this before embarking on the use of Part Five. This may well encourage the reader to tackle the whole of this excellent EEUA publication.

For ease of treatment the text has been grouped into chapters dealing with the preliminary choice of pump, inlet conditions, flow/head rating sequence, driver power rating, casing pressure rating, and sealing considerations. Each chapter is supported by a flow diagram providing a model of the sub-system of calculations. The tacit assumption underlying this arrangement is that the same full sophisticated and rigorous treatment will be applied to each and every pump installation. In practice, however, the treatment applied to the specification and selection of individual pumping installations can vary over a wide spectrum from a fairly cursory examination of a limited number of critical factors to the full and rigorous treatment assumed here. As outlined in Part One, this will depend on the criticality of each installation.

The flow diagram shown in Fig. 21.1 provides a model of the whole system of calculations to be carried out for centrifugal pumps. Each operation is cross-referenced to the relevant chapters or sections as appropriate, (numbers in right hand side of boxes). As explained in Part One, the system is broken down into three stages in order to cope with the needs of different users and/or different pump installations. The first stage covers the preliminary choice of pump type and speed; the second stage covers the more sophisticated treatment with particular reference to methods of control and corrections to the NPSH available; and the third stage covers information requirements for completion of the data sheet and for the needs of other related design work (i.e., civil, piping, electrical, instruments, etc.) which are dependent on the choice of pump type and speed.

For a very simple pumping installation such as a small cold water duty or duties with fluids having similar properties to cold water, with single stream operation (i.e., no parallel operation) an engineer might choose to use only stage I, leave out stage II as inappropriate, and rely on the pump manufacturer to supply the dependent information of stage III. This would assume that time was not a critical element for the release of information for layout, civil, piping, and electrical design purposes. Again, for a relatively simple pumping installation where the requirements were more critical for other aspects of plant design, an engineer may choose to omit stage II, and use only stage I followed up by a determination of the dependent information of stage III. Finally for a more complex installation with fluids other than cold water where NPSH available was much more critical, the full rigorous treatment would be applied, including the iterative confirmation of the selection of pump type and speed of stage II, before proceeding to stage III.

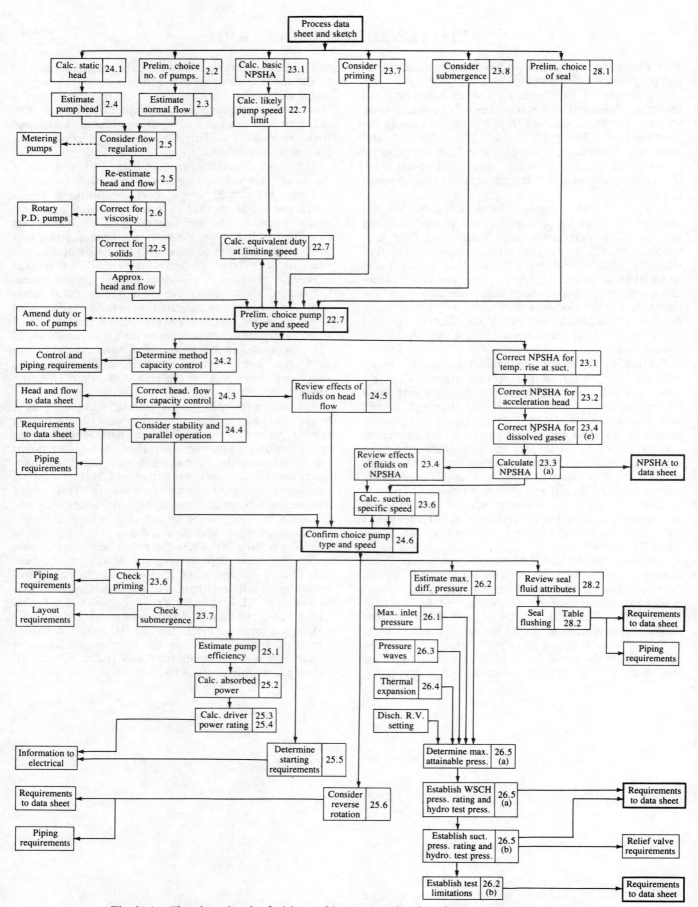

Fig. 21.1. Flowchart for the decision-making process for the selection of centrifugal pumps

CHAPTER 22. PRELIMINARY CHOICE OF PUMP

The process for the preliminary choice of pump for centrifugal pumps is modelled in Fig. 22.1.

22.1 Preliminary choice of number of pumps

For most current designs of process pumps with seals, the L.10 life is less than 8000 hours. Consequently pumps fall into reliability classes 4, 5, and 6 as defined in Appendix II with the short seal life accepted as the best 'state of the art'. The arrangement is one running pump rated at 100 per cent duty, with one identical pump as the spare either installed or available as a replacement, both having mechanical seals.

The only pumps known to be capable of Class 1 operation, are:

(a) glandless canned motor or canned coupling pumps;

(b) glandless wet stator motor pumps;

(c) cantilever shaft vertical immersed pumps;

(d) special purpose pumps designed to meet the mechanical seal needs, where the intrinsic L.10 life of the seal exceeds 20 000 hours.

Even these pumps will only achieve Class 1 operation under favourable operation and maintenance routines.

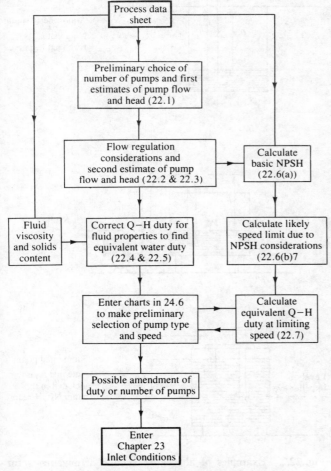

Fig. 22.1.   Preliminary choice of pump

22.2 Considerations of flow regulation

If $\dfrac{Q_0}{H_0} > 0.08$

then confirm a centrifugal pump as the first choice and:

(a) up to a flow of 3.5 l/s, select one running pump regulated by a control valve in the bypass line and take the second estimate of Q_0 to be 4 l/s;

(b) above a flow of 3.5 l/s through the pump, select regulation by throttle control;

(c) above a flow of 100 l/s through each pump check that the flow regulation range does not exceed 4 : 1, viz. 30 to 120 per cent of the nominal rating. If a pump is required to run for long periods at flows outside this range then revise the choice of the number of pumps or increase Q_0 by 17 per cent for the second estimate of Q_0 to allow for a permanent bypass.

22.3 Second estimate of pump head

If the preliminary choice is a centrifugal pump with throttle regulation then take the second estimate of pump head as

$$H_0 = 1.08H_s + 1.63H_r \qquad \text{(m)}$$

where
H_s = differential static head across the pump (see section 2.4)
H_r = total frictional head loss in the inlet and delivery system at normal flow

22.4 Effect of viscosity

Calculate

$$\phi = \frac{v}{Q_0^{1/2} \cdot H_0^{1/4}}$$

where
v = kinematic viscosity of the liquid at normal temperature (cSt)

If $\phi > 20$, then a rotary pump should be selected – see Part Four

If $\phi < 1$, ignore effect of viscosity

If $\phi > 1$, calculate the equivalent duty for a viscosity of 1 cSt as

$$Q = Q_0\left(1 + \frac{0.026}{Q_0^{5/6} \cdot H_0^{1/4}}\right)$$

$$H = H_0\left(1 + \frac{0.017}{Q_0^{5/6} \cdot H_0^{1/4}}\right)$$

22.5 Effect of suspended solids

For slurries of crystalline particles, first apply the corrections to Q_0 and H_0 for the viscosity of the carrier liquid, then apply the factor $(1 + 0.55x)$ to the differential head, where x is the fractional solids content by volume.

22.6 Influence of Net Positive Suction Head

(a) *Calculation of basic net positive suction head (NPSH)*

Basic NPSH

$$= h_{si} - h_{ri} + \frac{10.2}{\rho}\left(\frac{B}{1000} + P_i - P_v\right) \quad \text{(m)}$$

where

h_{si} = static liquid head over pump inlet (m)

This is measured from the lowest liquid level to the pump centre and is negative when the liquid level is below the pump centre. See Fig. 22.2. The lowest level is not necessarily the extreme value, it is often more economic to empty a vessel by a small auxiliary drain pump rather than specify an extraordinarily low NPSH requirement for the main pump.

B = Minimum barometric pressure at pump location (mbar)

Use 0.94 of the mean barometric pressure from local meteorological data to allow for weather extremes.

P_i = Minimum working value of the gas pressure on the free liquid surface in the inlet vessel (bar g)

This is negative for vessels under vacuum.

P_v = Vapour pressure of the liquid at the maximum operating temperature, including the correction for temperature rise in bypass system in section 23.2 (bar a)

When the fluid is complex – generally a hydrocarbon mixture – the 'bubble point' pressure is required, not the Reid vapour pressure nor pressure readings from the vapour space of the supply vessel.

h_{ri} = Frictional head loss in the inlet piping system (including the loss in the inlet strainer) calculated for *maximum* flow. Note that a typical pump control system will yield a maximum flow of 115 per cent of the normal flow. Higher maximum flows can arise – refer to section 23.3.4. If the basic NPSH is less than 3.25 m then verify h_{ri} by site pressure drop measurements or re-examination of the pressure drop calculation.

ρ = Liquid density at the maximum operating temperature. For slurries use the density of the *mixture* (kg/l)

Note that the effect of temperature on vapour pressure outweighs the effect of temperature on the density of the fluid.

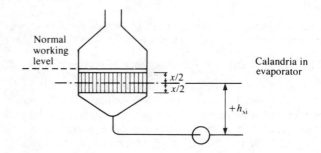

Calandria in evaporator

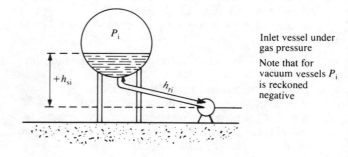

Inlet vessel under gas pressure

Note that for vacuum vessels P_i is reckoned negative

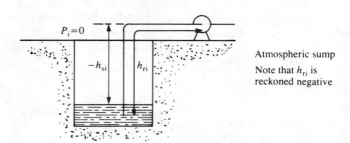

Atmospheric sump

Note that h_{ri} is reckoned negative

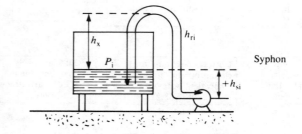

Syphon

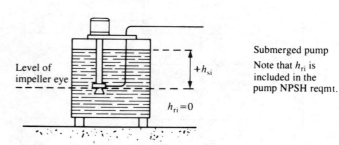

Submerged pump

Note that h_{ri} is included in the pump NPSH reqmt.

Fig. 22.2. Examples of alternative suction arrangements for centrifugal pumps

Notes

(a) For liquids at boiling point the function in the brackets is always zero so that the basic NPSH becomes simply

$$h_{si} - h_{ri} \quad \text{(m)}$$

However, when there is heat transfer the liquid is not in an equilibrium state. Then:

(i) For an internal heat source, e.g., a calandria in an evaporator, the column of liquid/vapour in the calandria tubes reduces the pressure at the calandria base and induces circulation. For pumps the effective quiescent liquid surface is then very nearly the midpoint of the calandria – see Fig. 24.2.

(ii) For an external heat source, e.g., nominally refrigerated liquid entering a storage vessel, small differences in temperature may occur between the liquid and its vapour. Check that such liquid entry connections are above the maximum liquid level to permit 'flashing' and thus closely approach temperature equilibrium.

(b) For small vessels, where the residence time for the fluid is less than 100 seconds, the turbulence can be high and the effective static head less than the geometric elevation. The effect is not accurately predictable. As a rough guide, take

$$\text{effective } h_{si} = \text{geometric } h_{si} - \frac{50}{\lambda \cdot v^{1/3}} \quad \text{(m)}$$

where λ is as defined in section 23.7(c).

An allied effect arises from swirl induced by rotary mechanical agitators. Fortunately its magnitude is sufficiently small to be ignored except for high agitation intensities (>1 kW/te) in systems where the pump NPSH is critical. If satisfactory experience with a similar installation is not available, then model tests should be considered.

(b) *Primary values of NPSH*

At constant pump speed, the NPSH required by pumps increases with capacity. The zonal boundaries in the generalised pump selection charts (given later) correspond approximately to a basic NPSH of 10.5 m.

If the calculated value of the basic NPSH is less than 10.5 m then the likely limit on the rotational speed of the pump impeller is given by

$$N_{max} = \frac{100(\text{NPSH}_{basic} - 1)^{3/4}}{Q^{1/2}}$$

22.7 Preliminary choice of pump type and speed

Using the corrected values of flow and head from sections 22.4 or 22.5, enter the charts given in section 24.6.

The speed thus chosen may exceed the limit given in section 22.6. If it does, then obtain the equivalent flow/head duty at the speed N_{max} and re-enter the charts.

These equivalent duties are

$$Q^* = Q \left| \frac{N}{N_{max}} \right|$$

$$H^* = H \left| \frac{N}{N_{max}} \right|^2$$

This step assumes that a belt or gear drive is acceptable if the equivalent duty falls in the zone chosen originally. If it falls in a zone calling for a higher pump speed then consider special pumps with inducers or double-entry impellers, or the arrangements discussed in section 24.6(e).

CHAPTER 23. INLET CONDITIONS

The process for confirming that the inlet conditions are suitable for the preliminary choice of pump are modelled in Fig.23.1.

23.1 Correction to basic NPSH for temperature rise at pump inlet

(a)   All pumps have internal leakage flows which constitute an equivalent bypass system; these are taken into account on pump performance curves given by the manufacturers. All pumps that can briefly operate against a closed throttle control valve with nominally zero flow through the pump should be allowed a NPSH correction of 0.5 m to the value calculated for maximum flow.

(b)   An alternative correction applies to pumps where the bypass line returns direct to the pump inlet branch instead of the supply vessel, and the control valve is in the delivery line.

FLOWCHART

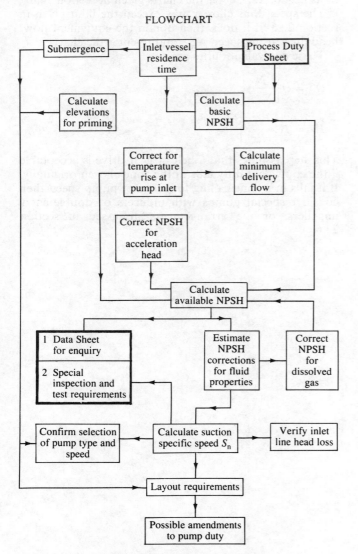

Fig. 23.1.   Inlet conditions

Ignoring radiation and convection losses or gains

$$\Delta T_i = \frac{H}{102 \cdot \xi} \cdot \frac{100}{\eta_1} \cdot \frac{q_b}{q_d}$$

where

ΔT_i = rise in temperature of liquid at pump inlet (°C)

η_1 = pump hydraulic efficiency at point (Q, H) (per cent)

H = differential head across pump (m)

q_d = minimum continuous flow to delivery point (l/sec)

q_b = corresponding flow through bypass (l/sec)

Q = corresponding flow through the pump (l/sec)

ξ = liquid specific heat (kJ/kg °C)

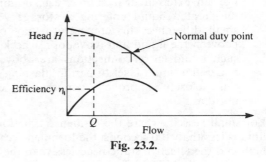

Fig. 23.2.

Recalculate the basic NPSH using the vapour pressure corresponding to the specified inlet temperature plus the increment ΔT_i, and the frictional head loss (h_{ri}) in the inlet system for flow q_d, not the maximum flow through the pump.

23.2 Correction to the basic NPSH for acceleration head

If the flow through the pump is automatically controlled with a narrow proportional band or fast integral action then an allowance should be made for the head required to accelerate the liquid in the inlet line when the control valve operates, in order to avoid relaxation fluctuations in flow.

Take this correction as 0.5 m for systems where the total length of line from the inlet vessel branch to the pump is less than $D/15$ metres (D = pipe diameter in mm).

(a)   For longer lines, if the system control characteristic is nearly linear

$$h_a = \frac{\sum (v \cdot l)}{10t}$$

where

h_a = acceleration head (to be subtracted from the basic NPSH) (m)

v = velocity in inlet line section at normal flow (m/sec)

l = length of inlet line section in which liquid velocity is v m/s (m)

t = time for the control valve to operate from shut to full open (sec)

Determine the value appropriate to any specific system: as a rough guide assume a preliminary value of $D/24$ seconds, where D is the nominal pipe diameter in mm. Consult a control specialist if necessary.

Note that the time for the control valve to operate from open to shut can be ignored.

Where both inlet and delivery lines are long, the acceleration head applicable to the inlet line is limited, so that

$$h_a = \frac{\sum (v \cdot l)}{\sum (V \cdot L) + \sum (v \cdot l)} \cdot (H - H_s)$$

where V and L refer to the delivery line, H is the differential head across the pump at normal flow, and H_s is the differential static head across the pump.

23.3 Calculate available NPSH

(*a*) Calculate the basic NPSH as in section 22.6 then subtract either 1 m or the corrections given in sections 23.1 and 23.2 if they total more than 1 m. Apply the correction for dissolved gas given in section 23.4(e). The NPSH value so obtained corresponds to *maximum* flow through the pumps.

However, pumps are usually ordered for the *normal* flow duty. Near the best efficiency point the NPSH required by a pump roughly varies as flow, so that in specifying the *normal* flow duty only 80 per cent of the NPSH value calculated above should be inserted on the Pump Data Sheet. Thus for most pump duties the available NPSH given to the pump supplier is

$$\text{NPSH} = 0.8(\text{NPSH}_{\text{basic}} - 1) \qquad \text{(m)}$$

Exceptions to this rule are:

(i) pumps operating with available NPSH less than 1.8 metres as given by the above expression;

(ii) pumps having a suction specific speed above 0.4 calculated as section 23.5;

(iii) pumps where the maximum flow is specified, or exceeds 115 per cent of normal flow. This particularly applies to pumps regulated by a control valve in the bypass line, where the working point on the QH curve ranges between normal and maximum flow through the pump.

For these exceptions, specify the duty at maximum flow through the pump and the available NPSH at this flow, in addition to the rated duty point.

(*b*) Apply the corrections for fluid properties as section 23.4 then calculate the suction specific speed as section 23.5 to assess pump sensitivity to inlet conditions. Do not insert this corrected NPSH in the pump data sheet.

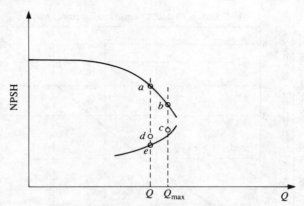

Fig. 23.3. Definitions of NPSHA

Point	Definition
a	**Basic NPSHA at rated flow**
b	**Basic NPSHA at maximum flow**
c	**Corrected NPSHA at maximum flow**
d	**80 per cent of the corrected NPSHA at point *c***
e	**Typical NPSHR from pump maker curve**

Note that the NPSH difference between *b* and *c* is the margin applied in this publication, whilst the difference between *d* and *e* represents the pump maker's margin, if any. Thus *ae* is the conventional margin between the NPSHA value at rated flow and the maker's value for NPSHR

23.4 Corrections to NPSH for fluid properties

The following corrections are mainly for discussions on layout and control, and with the exception of section 23.4(e) would not normally be disclosed to the supplier.

(*a*) *Suspended solids content*

The behaviour with slurries is notoriously difficult to predict; the only reliable test is to run the pump on the plant. For suspensions of hard crystalline solids, use the following empirical rules as a guide.

(i) Very fine suspensions where all the particles are less than 3 microns are best treated as viscous liquids – see Fig. 23.4.

(ii) For slurries having a range of particle sizes take

$$\text{NPSH}_{\text{slurry}} = \left(\frac{1}{1 + x}\right)\text{NPSH}_{\text{carrier}}$$

where x = fractional content by volume of solid particles and the NPSH for the carrier liquid has already been corrected for viscosity.

(*b*) *Viscosity*

An increase in viscosity increases the NPSH required to maintain performance – see Fig. 23.5 for the factor to apply.

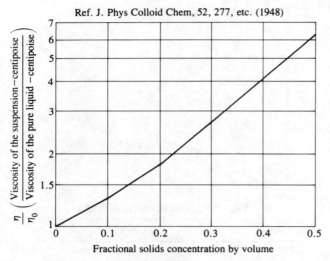

Ref. J. Phys Colloid Chem, 52, 277, etc. (1948)

Fig. 23.4.

(c) Thermal properties

The fluid thermal properties modify the pump response to NPSH conditions. Provided the liquid is stable and non-corrosive it is sometimes possible to operate at an NPSH less than the limit determined by tests on cold water.

(d) Occluded gas

No correction need be applied to the basic NPSH but it is *essential* that the layout is checked as in section 23.5(c).

(e) Dissolved gas

Where the inlet vessel is a gas scrubber the liquid is likely to be saturated with the gas. For such duties the

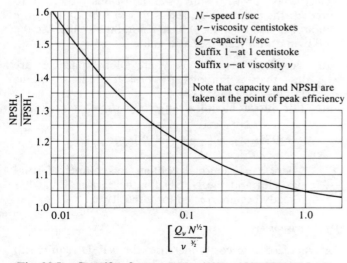

Fig. 23.5. Centrifugal pumps – variation of NPSH with viscosity

calculated basic NPSH may be deceptively high; the pump performance being affected by gas release rather than by the vapour release upon cavitation. The audible and destructive effects of cavitation are mitigated by the presence of the gas but a correction to the basic NPSH is required to avoid a shortfall in pump performance.

Apply the following procedure.

(i)   Use gas solubility data to construct the graph giving the relation between absolute pressure and the gas release in actual m^3/m^3 liquid. Gas release is normally insensitive to temperature: exceptionally some systems may need evaluation both at maximum and minimum pumping temperatures. Note that boiling liquids can be assumed to contain no dissolved gas.

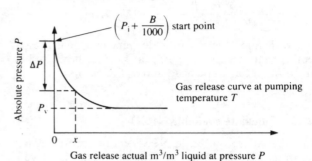

Fig. 23.6.   Gas release/pressure rating

(ii)   Calculate the stage specific speed N_s at the pump BEP.

(iii) From the gas release/pressure relation obtain the pressure reduction (ΔP) to point X (see Fig. 23.5).

Once formed, bubbles redissolve slowly and thus persist through the pump. The effect of such occluded bubbles depends on the stage specific speed, giving a significant performance loss when the bubble volume is $(2 + 20N_s)$ per cent. Only a small fraction of the liquid flow passes through the low pressure zone close to the impeller vanes where bubbles can be generated. Consequently the first estimate of X is taken as

$$(2 + 20N_s) \quad (m^3 \text{ gas}/m^3 \text{ liquid})$$

(iv) Recalculate the basic NPSH in section 22.6 replacing

$$\left(P_i + \frac{B}{1000} - P_v \right) \quad \text{by} \quad \Delta P$$

Note that this step assumes that the gas content gives a pseudo vapour pressure of

$$\left(P_i + \frac{B}{1000} - \Delta P \right)$$

(v)   Obtain the final value of the basic NPSH by subtracting the following correction from this recalculated basic NPSH

$$\frac{10}{\rho} (S_n - 0.12)^2 \left(\frac{B}{1000} + P_i - P_v - \Delta P \right) \quad \text{(m)}$$

where

$$S_n > 0.12$$

23.5 Calculation of suction specific speed

$$S_n = \frac{N \cdot Q^{1/2}}{K \cdot H_n^{3/4}}$$

where

S_n = suction specific speed as a dimensionless number

K = constant, numerically 175 including g at 9.81 m/s² for the following units:

N = impeller speed (r/sec)

Q = normal flow through pump (l/sec)

For pumps having double-entry impellers take this value as

$$\frac{Q}{2\left(1 - \dfrac{N}{6Q}\right)} \quad \text{where } \frac{N}{Q} < 3$$

H_n = net positive suction head, taken as the value to be expected for a 3 per cent head drop, over the first impeller only, on a pump test with cold water by correcting the available NPSH for the effect of the process liquid properties (m).

The suction specific speed is a guide to layout and pump specification as follows.

(a) $S_n < 0.12$

Below this value pumps are essentially free from cavitation. This limit should be applied for pumps handling very corrosive gas free liquids where the ordinary materials of construction which are resistant to cavitation attack cannot be used. It should also be applied for pumps sited in 'quiet' areas where the noise energy generated by the pump is important.

For less critical duties an acceptable life expectancy is obtained at higher values of S_n as follows.

(b) $0.12 < S_n < 0.4$

An inlet reducer is not essential, the inlet pipe bore (D) can equal the bore of the pump inlet connection. Bends directly connected to the pump should have a radius exceeding $5D$.

(c) $0.4 < S_n < 0.7$

This range is reached by the best available commercial pumps when handling nominally clean liquids. Because the pump operates continuously with incipient cavitation, the material of construction should be resistant to corrosion.

The operational flow range is restricted to 40 to 110 per cent of the pump capacity at the best efficiency point.

Check the line diagram/layout to ensure that the inlet pipe is straight and free from obstruction (e.g., Tee-piece or thermocouple sheath) for at least three diameters ahead of the pump inlet. Where an inlet reducer is provided, it should have a total cone angle less than 30 degrees and be installed directly at the pump inlet. Verify the pressure drop calculations for the inlet system.

(d) $S_n > 0.7$

This can only be reached by special pumps employing an inducer and dealing with clean non-corrosive liquids.

Check the proposed layout to ensure that the pumps are adjacent to the supply vessel. Inlet isolation valves should be line-size and of the ball or other full-flow type. A conical reducer is required with the optimum total cone angle of 16 degrees, installed directly at the pump inlet.

The inlet pipe should be straight and free from obstruction for at least five pipe diameters ahead of the inlet reducer.

The permitted operational flow range is restricted to 70 to 105 per cent of the pump capacity at the best efficiency point.

Verify the pressure drop calculations for the inlet system.

23.6 Priming

(a) Simple centrifugal pumps will not work unless both pump and inlet lines are full of liquid, i.e., primed.

The static liquid level at pump inlet (H_{si}) is the height to which the liquid will rise above pump centre.

$$H_{si} = h_{si} + \frac{10.2 \times P_i}{\rho_m} - h_{ri} \quad \text{(m)}$$

where

h_{ri} = frictional loss in inlet system at normal flow (m)

Note that this term is included because general operational policy requires that the standby pump can start whilst the operating pump continues to run.

h_{si} = the difference in elevation between the pump centre and the lowest working level in the inlet supply vessel (m)

Note that h_{si} is negative when the liquid level is below the pump centre.

P_i = minimum working gas pressure in the inlet vessel (bar g)

Note that P_i is negative for vessels under vacuum, or taken as zero if the priming vent is piped to the gas space of the inlet vessel.

ρ_m = maximum liquid density (kg/l)

(b) Whenever possible ensure priming by bringing the static liquid level above the highest point of the pump casing (not the pump centre line). In this case check the line diagram to ensure that:

(i) the vent is shown;
(ii) the vent is piped to a safe disposal point when the pump is handling a hazardous liquid;
(iii) when the inlet vessel is working under vacuum, the vent is piped to the gas space of the inlet vessel;
(iv) when the vent is piped, then a visible indicator is fitted in the vent line adjacent to the pump.

(c) If the static liquid level is unavoidably below the highest point of the pump casing, then use one of the following schemes.

(i) A self-priming pump
(ii) An external priming arrangement for standard centrifugal pumps. Check that the line diagram then shows:

(1) a non-return valve (termed a foot valve) situated in the inlet line below the minimum liquid level;
(2) a method of supplying liquid to fill the pump.

(iii) A standard pump installed in a pit, if the pumped liquid is not flammable.
(iv) A vertical wet-sump pump.

Check the proposed layout to see whether the inlet line rises at any point to form a syphon. Syphons should be avoided, except as a substitute for an inlet isolation valve for particularly aggressive liquids.

23.7 Submergence

Submergence is the depth of the inlet opening beneath the free liquid surface. If the submergence is too small there is a risk of vortex formation and gas entrainment.

(a) Check the proposed layout against the acceptable arrangements shown in Fig. 23.7 using the liquid velocity corresponding to *maximum* flow.

If the liquid velocity should be decreased to meet the available submergence then a cone reducer at the vessel exit branch may be used. Note that dimension 'D' in Fig. 23.7 is the diameter at the entry to such a reducer.

The effect of liquid viscosity is pronounced but not reliably predictable. As a rough estimate take

$$V_c = 0.15(v^{1/3} - 1) + V_w$$

where
V_w = critical inlet velocity for a given submergence in cold water (m/sec)

v = kinematic viscosity of liquid at operating temperature (when it exceeds unity) (cSt)
V_c = critical inlet velocity for the same submergence in the liquid (m/sec)

(b) The arrangements shown in Fig. 23.7 assume that the vessel or sump is large. For pumps drawing from small tanks, the submergence should be sufficient to suppress the effects of the fresh liquid which enters the tank to maintain the liquid level.

(c) A useful parameter is the residence time, λ, defined as the time taken in seconds at constant maximum flow to pump the volume between the effective undisturbed level and the topmost point of the vessel exit branch.

(i) The residence time λ should not be less than 100 seconds without satisfactory operational experience from an identical system.
(ii) When $100 < \lambda < 200$ increase the calculated minimum submergence by the factor

$$\left(\frac{200}{\lambda}\right)^2$$

(iii) When $\lambda > 200$, the effect of the vessel size is neglected.

(d) Check the method of introducing the supply liquid into the tank:

(i) For $\lambda < 200$, the supply is best admitted over a weir having a suppressed nappe.
(ii) For $\lambda > 200$, a simple pipe connection is satisfactory provided that the maximum inlet velocity is 0.6 m/s and arrangements embodying deliberate swirl are avoided.
(iii) In packed towers the liquid is assumed to rain uniformly on the reservoir surface. Check that the froth is allowed sufficient time for gas disengagement before the liquid enters the pump inlet line.

(e) Check the proposed hydraulic design.

(i) If swirl is deliberately induced, the proposed layout should be compared with a geometrically similar layout already known to work satisfactorily. The scale critical submergence will then be nearly correct for equal inlet velocities.
(ii) Large vertical canned (caisson) pumps and vessels with tangential inlet or central bottom exit invariably require a vortex suppressor. Vortex suppressors are difficult to design except by model tests; simple crosses in exit lines from vessels are ineffective.

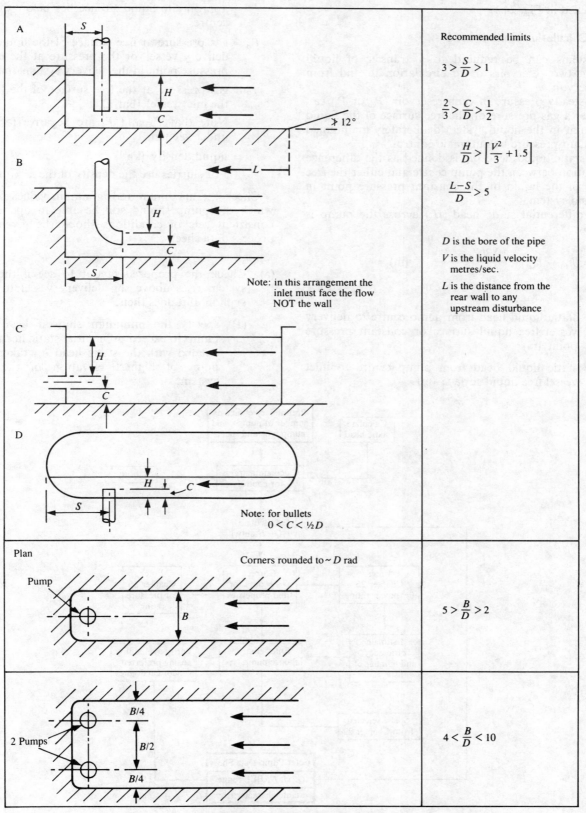

Fig. 23.7. Submergence limits

CHAPTER 24. FLOW AND HEAD RATING

The process for the specification of flow and head rating is modelled in Fig. 24.1.

24.1 Calculation of static head

All systems can be reduced to a transfer of liquid between two reservoirs or to circulation to and from one reservoir.

The steady pressure in each reservoir (P_d or P_i) can be either a gas pressure on the free surface of the liquid or a point in the liquid system deliberately maintained at constant pressure by automaic control.

The static liquid head (h_s) is defined as the difference in elevation between the pump centre and either the free surface of the liquid or the constant pressure point in the liquid system.

The differential static head (H_s) across the pump is given by

$$H_s = (h_{sd} - h_{si}) + \frac{10.2(P_d - P_i)}{\rho} \quad \text{(m)}$$

where

h_{sd} = static liquid head from pump centre to delivery vessel free liquid surface or constant pressure point (m)

h_{si} = static liquid head from pump centre to inlet vessel free liquid surface (m)

Note that h_{sd} and h_{si} are negative when the appropriate liquid surface lies below the pump centre.

P_d = gas pressure at free surface of the liquid in the delivery vessel, or the pressure at the constant pressure point in the delivery system (bar g)

P_i = gas pressure at the free surface of the liquid in the inlet vessel (bar g)

Note that P_d and P_i are negative for vessels under vacuum.

ρ = liquid density (kg/l)

For slurries use the density of the mixture.

Check the maximum and minimum values of H_s for which the pump should still be able to operate. Self-consistent sets of conditions should be given in the process data sheet.

Notes

(a) Check the proposed layout to see if the piping system rises above the delivery vessel to form a syphon. If it does then:

(1) specify the minimum shut-off head of the pump to exceed differential static head H_s calculated with the static head h_{sd} taken to the point of highest elevation of the piping system;

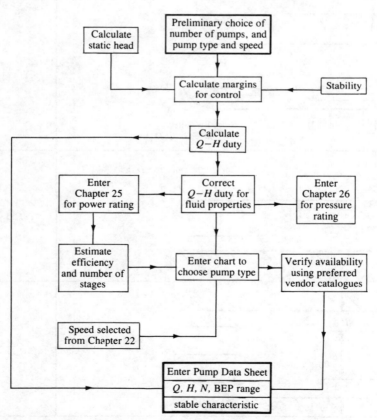

Fig. 24.1. Flow/head rating sequence

(2)   specify a stable pump characteristic.

Syphons on the inlet piping system can be ignored because a separate priming arrangement is needed before the pump can start.

(b)   For booster or circulating pumps P_d and P_i increase or decrease together so that $(P_d - P_i)$ is the proper parameter. For such duties do *not* take the differential pressure as $(P_{dmax} - P_{imin})$.

24.2   Calculation of margin for control

(a)   *Choice of control method*

The control system is normally determined by process requirements, but the pump characteristics may affect the choice of throttle, bypass, or speed regulation.

Control is not to be construed only as automatic control; the control margin is still required if the pump is to be regulated by an operator using a manual valve. The margin can only be reduced if the range of control is reduced.

The methods of control are as follows.

(i)   Flow throttling by a series control valve in the discharge line from the pump (Fig. 24.2). This is the common system, and should be adopted whenever possible, as it is simple and reasonably efficient.

(ii)   Bypassing a portion of the pump flow to control forward flow or to maintain a minimum flow through the pump at all times (Fig. 24.3). In the first case the bypass control valve is normally modulating, in the other it is normally closed.

As a guide a bypass system is required when

$$Q < 0.3 \cdot H^{1/2} \quad \text{or} \quad Q < 3.5$$

where

Q = flow through pump (l/sec)
H = differential head (m)

Bypass systems are often required for:

(a)   Slurry pumps (when variable speed drive is excluded). This often takes the form of a pump delivering to an overhead tank which overflows the surplus back to the supply tank.
(b)   Peripheral pumps.

Bypass systems may be required:

(c)   When the delivery route may be closed off for an extended period (a pump trip is an alternative).
(d)   To bring the operating point within the flow range given by section 24.3(*d*).

In these cases it may be sufficient to use throttle control together with a permanent bypass flow through a restrictor (Fig. 24.4).

(iii)   Speed variation (see Fig. 24.5). The cost of variable speed drives is high and they are considerably

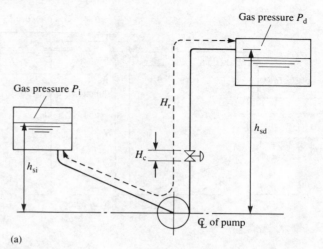

(a)

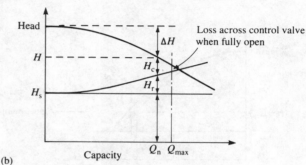

(b)

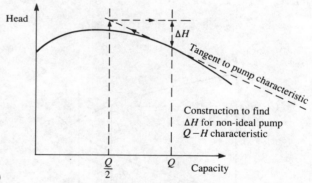

(c)

Fig. 24.2.   Throttle control

more complex both in terms of operation and maintenance so they are only justified for:

(a)   large pumps (consuming more than 200 kW) when the flow is required to vary continually;
(b)   pumps dealing with abrasive slurries;
(c)   pumps operating for significant periods at low flows and in systems with high control valve pressure drop.

Note that variable speed regulation is prone to instability when operating below 30 per cent capacity and is therefore unsuitable for level control in vessels unless supplemented by a secondary control system.

In the special case of axial or cone (mixed) flow

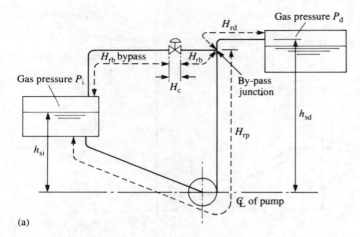

(a)

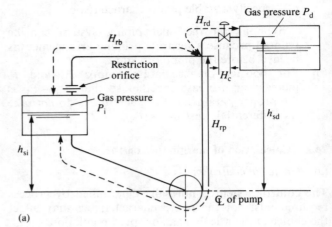

(a)

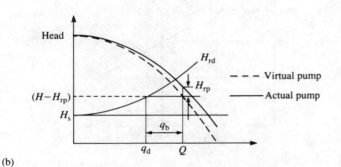

(b)

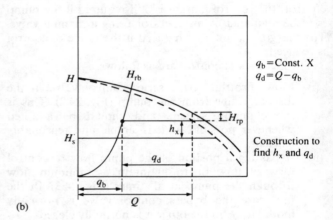

(b)

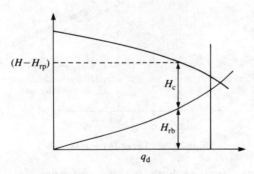

(c)

Fig. 24.3.   Bypass control

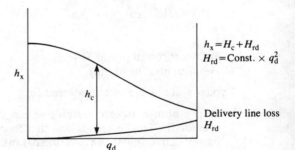

(c)

Fig. 24.4.   Throttle control plus bypass orifice

pumps on circulating duties (where the differential head is not accurately calculable) consider a belt drive so that the speed can be easily adjusted by a change of pulleys.

(b)   Throttle control

The control valve pressure drop is related to the pump system resistance by parameter r, defined by

$$r = \frac{H_c}{\Delta H + H_r}$$

where

H_c = control valve head loss (m)
ΔH = virtual head loss in the pump (m) (see Fig. 24.3(b))
H_r = flow dependent head loss in the system, excluding the control valve (m)

For adequate control in most installations, $r = 0.5$ at normal flow. Then

$$H_c \sim 0.5(\Delta H + H_r) \text{ at normal flow}$$

Normal flow is defined as the largest process flow

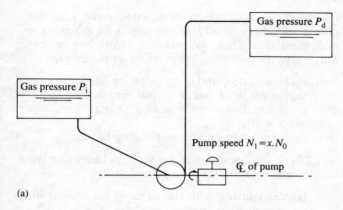

Gas pressure P_d

Gas pressure P_i

Pump speed $N_1 = x.N_0$

$\mathbb{C}$ of pump

(a)

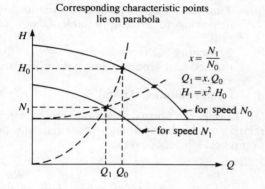

Corresponding characteristic points
lie on parabola

H

H_0

N_1

$x = \dfrac{N_1}{N_0}$

$Q_1 = x.Q_0$

$H_1 = x^2.H_0$

— for speed N_0

— for speed N_1

Q_1 Q_0

Q

(b)

Fig. 24.5. Variable speed regulation

required for plant operation. This should be distinguished from the maximum flow through the pump obtained when the control valve is fully open.

When reviewing an existing pump installation the appropriate value of ΔH is obtained by the construction shown in Fig. 24.3(c).

The typical value of the virtual head loss in a pump is 15 per cent of the pump total differential head. Therefore, for adequate control at normal flow

$$H_c = 0.63H_{r,\,normal} + 0.08H_{s,\,max} \quad \text{(m)}$$

This design control valve loss will change when:

(i) a rise to the shut-off of more than 15 per cent of the differential head at duty point is required for pump stability;

(ii) the maximum flow exceeds 115 per cent Q_0;
(In this case the pump duty is based on the maximum flow. The control valve head loss when fully open is taken as

$$H_{co} = 0.19H_{r,\,max} + 0.03H_{s,\,max} \quad \text{(m)}$$

where $H_{r,\,max}$ is the system frictional loss at maximum flow. Alternatively, it may be more convenient to use a notional normal flow defined as (Maximum flow/1.15))

(iii) the pump had H is large ($H > 800$ m) or the pump power is large ($E > 1$ MW) and the control range of flow exceeds $3:1$. Then obtain r from Fig. 24.6 and adjust the calculated value of h_c by multiplying by the factor $2r$.

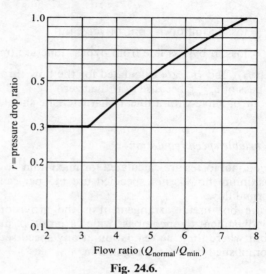

Flow ratio ($Q_{normal}/Q_{min.}$)

Fig. 24.6.

(c) Bypass control

This is a more difficult problem than throttle control; the following empirical rules are not accurate and a check should be made using the actual pump characteristic.

Assume that the normal bypass flow (q_b) is 17 per cent of the normal delivery flow (q_d), so that the normal flow through the pump is 117 per cent of q_d.

Unless a larger value of maximum flow to the delivery point is required

$$q_{d,\,max} = q_{d,\,normal} \cdot \frac{Z + 0.21}{Z + 0.18}\left(\frac{\Delta H_s}{0.21H} + 1\right)^{0.4}$$

where

$$Z = \frac{H_{rd}}{H_{s,\,min}} \quad \text{(m)}$$

$$\Delta H_s = H_{s,\,max} - H_{s,\,min} \quad \text{(m)}$$

$$H = H_{s,\,max} + H_{rd} + H_{rp} \quad \text{(m)}$$

H_{rp} = frictional loss in the pump inlet and delivery lines up to the bypass junction, at normal flow (m)

With this system the maximum flow through the pump occurs at shut off flow to the delivery point, when the control valve is full open, and is given by

$$q_{p,\,max} = q_{d,\,max} \cdot \left(\frac{Z + X + 0.15}{(Z+1)(X+0.15)}\right)^{0.4}$$

where

$$X = \frac{H_{rp}}{H}$$

The maximum permissible friction loss in the bypass line, and the size of the control valve, is fixed by the condition of delivery flow shut-off. Then

$$(H_{co} + H_{rb}) < H_{s, min}$$

where

H_{co} = loss in full-open control valve (m)

H_{rb} = loss in bypass line from bypass junction (m)

Both H_{co} and H_{rb} are calculated for the flow through the bypass of $q_{p, max}$ because q_d is then zero.

A control valve with a linear characteristic should be used.

(d) Variable speed regulation

The frictional losses are calculated for maximum flow. If the maximum flow is not specified use 115 per cent of the normal flow.

A rare but useful arrangement is the provision of speed adjustment to encompass wide variation of pump duty but where the true regulation at any speed setting is accomplished by throttle regulation.

24.3 Calculation of Q–H duty

(a) For throttle control

$$H = H_s + H_r + H_c$$

where

H = differential head across pump (m)

H_s = differential static head (m)

H_r = total frictional head loss in inlet and delivery lines (m)

H_c = head loss across control valve (m)

For bypass control with the control valve in the bypass line and for the variable speed control

$$H = H_s + H_r$$

Check that H_r is calculated for the correct flow.

(b) Add the following cumulative margins to the basic differential head H

Pump or duty classification	Margin (%)
Pumps handling clean liquids	Nil
Pumps handling nominally clean liquids	2
Slurry pumps	7
Magma pumps	10
Pumps running at variable speed	Nil
Pump duties where $H_r < 0.14H_s$	2
Duties with gas occlusion	See section 24.5(b)

Common practice is to construct pumps in corrosion resistant materials, consequently no design margin is added on this account.

(c)
Pump manufacturers guarantee only one duty point; all others should be expressed as minimum or maximum quantities. Where the pump has multiple duties give the following information on the enquiry:

(i)   maximum flow and corresponding head;
(ii) maximum head and corresponding flow;
(iii) head and flow corresponding to maximum value of $(Q \times H)$;
(iv) minimum flow and corresponding head.

(d)   Check the line diagram for systems likely to have a high maximum flow, for example:

(i)   Bypass control, with the valve in the bypass line, set to control at constant pressure. The pump flow is large when handling a liquid of higher density than that used to calculate the duty differential head.
(ii) Transfer pumps without automatic flow control delivering to an empty tank from a full supply tank.
(iii) Branching delivery systems where the flow is divided and separate flow controllers are not provided.
(iv) Pumps running in parallel.

(e)   Specify the normal flow points to be within a range relative to the pump design point (i.e., the flow at the pump best efficiency point).

Pump or duty classification	Range as % BEP flow
Peripheral pumps	20–100
Low NPSH duties where $0.4 < S_n < 0.7$	70–90
Low NPSH duties where $S_n > 0.7$	85–95
Pumps required to have zero cavitation	85–95
Fixed speed pumps with bypass control valve	25–75
Variable speed slurry pumps	85–100
Slurry pumps with minimum attrition	90–110
Pumps handling nominally clean liquids on throttle control	50–100
Self priming pumps	20–115
Duties where $H_r < 0.14H_s$	75–110

(f)   For pumps with variable speed drive of the fluid or magnetic slip coupling type specify the normal speed (for normal flow) as a fraction f of the maximum speed where

$$f = 0.88 + 0.12\left(\frac{H_{sm}}{H}\right)$$

H_{sm} = minimum differential static head (m)

H = differential head including margins (m)

24.4 Stability and parallel operation

(a) Stability

A stable Q–H characteristic is defined as one where the differential head H decreases continuously from the shut-off point as the flow increases.

The shut-off head should exceed 120 per cent of the differential head at the best efficiency point of the pump for the following duties:

(i) when the pump has a bypass control system with the control valve in the bypass line;

(ii) when the pump has a throttle or variable speed control system; and also

$$H_r < 0.14 H_{s,min}$$

where

H_r = total frictional loss in the inlet and delivery piping systems at normal flow (m)

$H_{s,min}$ = rated differential static head across the pump (m)

(b) Parallel operation

When two or more pumps run continuously in parallel, specify all the pumps to have:

(i) a stable characteristic;

(ii) identical shut-off heads to within a tolerance of 0 to +20 per cent of the rise to shut off from the duty point;

(iii) a driver rated for the limiting power demand with maximum flow. It is desirable for the pump to have a non-overloading power/capacity characteristic, otherwise the driver rating should cover the maximum power demand of one pump operating alone – see Fig. 24.7.

(c)

For parallel operation check that the engineering line diagram shows:

(i) a non-return valve at each pump discharge;

(ii) For systems defined in section 24.4(*a*)(ii) a throttle control valve in the discharge of each pump, *not* a single control valve in the common delivery line.

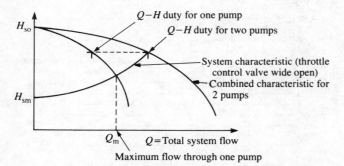

Fig. 24.7. Pumps in parallel operation

(d)

A useful rough estimate of the maximum flow through one pump (Q_m) of a pair is

$$Q_m = \frac{1.1Q}{\left(1 - \dfrac{0.75 H_r}{H_{so} - H_{sm}}\right)^{0.4}} \quad \text{(l/s)}$$

where

Q = rated capacity of one pump (l/s)

H_{so} = shut-off head, taken initially as 115 per cent of the rated pump differential head (m)

H_{sm} = minimum differential static head (m)

H_r = total system friction loss including the loss in the throttle control valve when *wide open* (i.e., at flowrate 1.1Q) (m)

Redetermine Q_m when the actual pump characteristic is available.

24.5 Corrections to Q–H duty for fluid properties

The following corrections are for use in design discussions and would not normally be part of the specification to the pump supplier.

Where any fluid property varies significantly, the extreme value should be the base for a set of self-consistent conditions supplementing the original sets of process conditions.

(a) Viscosity

(i) The pump capacity and differential head are not significantly affected by viscosity provided

$$v < Q^{1/2} \cdot H^{1/4}$$

where

Q = normal flow through pump (l/sec)

H = differential head at flow Q (m)

v = kinematic viscosity of liquid at operating temperature (cSt)

(ii) The useful range of centrifugal pumps is limited by

$$100 \cdot Q^{1/2} \cdot H^{1/4} > v$$

(iii) For radial flow centrifugal pumps an estimate of the reductions in capacity, differential head and hydraulic efficiency is given in Fig. 24.8.

(b) Entrained gas

Above 50 per cent of the pump BEP design capacity the effect of entrained gas is roughly equivalent to a reduction in speed – see Fig. 24.9. Below 50 per cent capacity the effect is unpredictable.

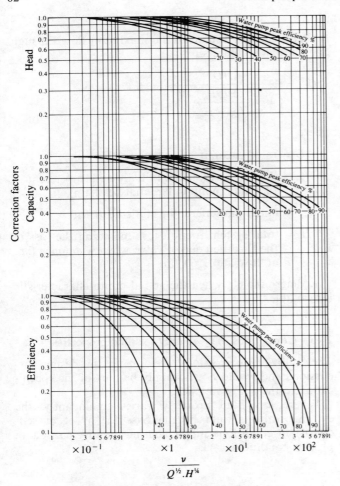

Fig. 24.8. Viscosity corrections

(iii) the suspended solid particle content is small, viz.

$$w < 100 \cdot \left(\frac{Q}{N}\right)^{1/3}$$

where

    w = solids content by weight (mg/kg)

    Q = normal flow through pump (l/sec)

    N = impeller speed (r/sec)

(*d*)

Slurries are typically particles of crystalline solid suspended in a liquid carrier. Slurries in a viscous carrier liquid are called 'magmas'. The behaviour of slurries is notoriously difficult to predict; the only reliable test is to run the pump on the plant. The following empirical rules should be regarded only as a guide.

(i) Very fine suspensions, where all the particles are less than 3 microns in size, can be treated as viscous liquids – see graph of equivalent viscosity, Fig. 23.4.

(ii) The dynamic effect of particles greater than 3 mm in size can be neglected provided the flow channels are large enough to avoid jamming, i.e., the least diameter of the channel should be more than three times the largest particle size.

(iii) For particles of intermediate size obtain the head/efficiency factor from Fig. 24.10. Divide the required differential head by this factor to obtain the equivalent water duty.

(*e*)

It is sometimes necessary to avoid degradation of the particles by attrition to ease filtration problems. Then:

(i) specify control by variable speed;

(ii) adjust the pump duty so that for one impeller

$$Q > 0.3 \cdot H^{3/2}$$

(*f*) *Non-Newtonian fluids*

Sols, gels, paper stock, and similar mixtures cannot be described generally and pumps can only be selected by reference to an existing plant handling the identical mixture.

(*c*) *Suspended solids*

A 'nominally clean' liquid means that

(i) the suspended solids are not intrinsically abrasive;

(ii) 99 per cent of the particles are less than 76 microns (µm) size;

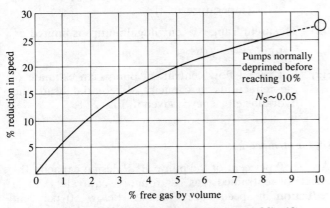

Fig. 24.9. Effect of entrained gas in pumped liquid

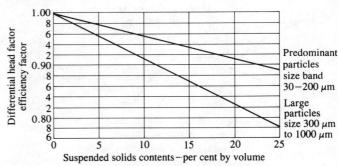

Fig. 24.10. Head/efficiency factor

(g) *Dissolved solids and gases*

These have no effect other than altering the density and the viscosity.

(h) *Compressibility*

For high head pumps a correction to the head may be necessary – see Note 1 of Appendix I.

(i) *Temperature sensitive liquids*

Some liquids are temperature-sensitive, e.g., polymerizing if the temperature exceeds a certain limit.

Check the minimum acceptable flow to the delivery point, remembering that the rise in temperature is given by

$$\Delta T = \frac{H}{102 \cdot \xi}\left(\frac{Q}{q_d} \cdot \frac{100}{\eta} - 1\right)$$

24.6   Guide to pump type and speed

(a)   *Single and 2-stage pumps; Table 24.1 (see Fig. 24.11)*

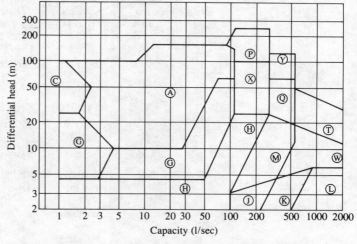

Fig. 24.11.   **Guide to pump type and speed – liquid viscosity 1 cSt**

Table 24.1

Zone Code	Number of stages	Impeller speed, N (r/sec)	Impeller type	Remarks
A	1	48	single entry radial	
C	1	48	peripheral	
H	1	16	single entry	
G	1	24	radial flow	
J	1	24		
K	1	16	axial propeller	horizontal
L	1	N		take $N < 500/Q^{1/2}$
M	1	As Zone Q or T, alternatively a single entry mixed flow (cone flow) impeller at same speed		
W	1	N	single entry cone flow	vertical take $N < 370/Q^{1/2}$
P	1	49		
Q	1	16.3	double entry radial flow	horizontal
X	1	24.5		
Y	2	24.5	double entry radial flow	2nd impeller may be single-entry
T	1	N	double entry	take $N < 500/Q^{1/2}$

This guide applies only for clean liquids of viscosity 1 cSt. For liquids of higher viscosity first obtain the equivalent duty.

Note that the inlet NPSH often determines pump speed. The chart is then re-entered after obtaining the equivalent duty at the speed given in Table 24.1 (see Fig. 25.6(b)).

(b)   *High head pumps – Table 24.2 (see Fig. 24.12)*

Notes to Table 24.2

(1)   In zones D, E, and F2 take reciprocating pumps as the first choice.

(2)   See section 25.1(d) for estimation of number of stages.

(3)   Multistage pumps are also classified by casing type:

Normal pressure (bar g)	Casing type
>100	Barrel
70–100	Barrel or axially split
40–100	Axially split
40–60	Axially split or cell*
<40	Cell construction

* In this pressure range the cell type construction is limited to 7 stages.

Table 24.2

Zone Code	No. of stages	Speed of impellers (r/sec)	Impeller type	Remarks
D	2	48	Peripheral	Clean liquid essential
E	2	up to 80	Peripheral	Radial flow inlet booster impeller may be required, forming a 3-stage pump.
F1	1	up to 380	Barske	An inlet helical inducer may be required.
F2	2	2-shaft		
B	3–9	48	Radial flow	Alternatively as Zone F1
R	3–9	normal range: 48–90 feasible range: 90–305	Radial flow	Inlet booster pumps may be required to provide NPSH. Upper bound line represents: $N \cdot S^{3/4} = 700$ where N = speed in r/sec and S = No. of stages
S	3–12	48	Radial flow	

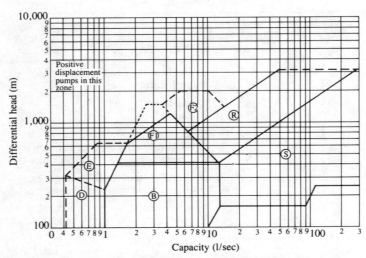

Fig. 24.12.   Guide to pump type and speed – pumps for clean liquids of viscosity 1 cSt above 100 m head

(c)   *Heavy slurry pumps using a carrier liquid of viscosity 1 cSt – Table 24.3 (see Fig. 24.13)*

These have a single overhung impeller. The graph gives the preferred speed in each zone appropriate for direct drive but a variable speed drive is recommended.

Notes to Table 24.3
(1) For a given configuration there is a critical fluid velocity above which the erosion rate increases rapidly.
(2) For carrier liquids having viscosities exceeding 1 cSt, first correct the pump duty using the approximate relationships given in section 22.5.

Table 24.3

Zone Code	Impeller speed (r/sec)	Impeller type	Remarks
A	48	Radial	Very short life expectancy
B	48	Radial	Alternatively use two pumps in series, each running at 24 r/sec
C	24	Radial	
D	–	Radial	Special pumps required
E, F, G & H	as on graph	Axial propeller	Pumps in this zone are normally employed on circulating duty.
M N	as on graph	Radial or Cone	

Table 24.4

Ref.	Type	Application	Remarks
1	Steam driven ejector	Aqueous cold solutions only	Very inefficient
	Compressed air driven ejector	Any service accepting aeration	
	Liquid driven ejectors	Slurries	
2	Air lift	Any service accepting aeration	Requires deep tube
3	Egging	Any service for a clean liquid	Intermittent action
4	Motor driven centrifugal pumps with volute chamber recirculation	Can be used for light slurries but then requires flushing through with clean liquid when employed intermittently	Capacity up to 80 l/sec Good efficiency
	Motor driven centrifugal pumps with liquid ring impeller action	Suitable only for clean liquids	Good performance at low NPSH

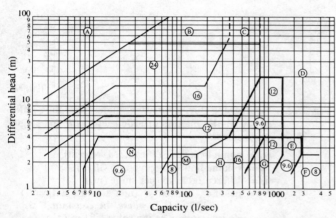

Fig. 24.13. Guide to pump type and speed – heavy slurry pumps at 50 Hz. Speeds given as r/sec

(d)   *Types of self-priming pumps*

(See Table 24.4.)

(e)   *Pumps for low NPSH service*

For some duties the pump selection guidance will conflict with the limit on pump rotational speed from NPSH considerations. If the service conditions cannot be altered then consider the use of:

(i)   Two pumps of 50 per cent capacity operating in parallel.

(ii)   Two pumps in series, the first being a low-speed booster providing sufficient NPSH for the main pump.

(iii)   A vertical canned (caisson) pump installed with the can below ground level so that the first impeller is provided with additional head. The depth of this first impeller is to be calculated on the basis that $S_n < 0.4$. The NPSH calculated at the top of inlet branch connection (not the centreline) should not be reduced below 2.5 velocity heads at maximum flow.

(iv)   A pumping pit, but note that this is normally forbidden for flammable liquids.

(v)   A super-cavitating pump to obtain values of S_n up to a maximum of 1.35.

(f)   *Pumps for deepwell service*

(i)   A vertical submerged pump with an integral electric motor driver, both sunk below the minimum liquid level. Use this scheme for clean water boreholes or for the recovery from cavern storage of low viscosity liquids ($v < Q^{1/2} \cdot H^{1/4}$).

Where the pumped liquid temperature exceeds 40°C a special investigation is required.

(ii)   A vertical submerged pump connected by a long vertical shaft to the driver mounted at grade. Consider for all services less than 300 m deep and more than 600 kW rated power.

This arrangement is also called a 'wet sump' pump and used in drain pit and sewage service for rated powers less than 600 kW.

CHAPTER 25. POWER RATING

The power rating sequence for the pump driver is modelled in Fig. 25.1.

25.1 Estimation of pump efficiency

(a) It is important to distinguish the definitions of efficiency.

(i) 'Pump efficiency' includes the effects of liquid viscosity, solids content, the frictional losses in the gland, capacity loss in gland flushing arrangements where the flushing liquid is taken from some point in the pump, and bearing losses.
This definition yields the power requirement at the shaft coupling to the driver or gearbox.

(ii) 'Overall efficiency' is the combination of the 'pump efficiency' and the 'motor efficiency', and the 'gear efficiency', if applicable.
This definition yields the power requirements at the input terminals of the electric motor and is useful where the motor is integral with the pump, e.g., glandless canned motor pumps.

(iii) 'Hydraulic efficiency' refers only to the intrinsic performance of the impeller/volute/casing combination.

The hydraulic efficiency is estimated from Fig. 25.2. For liquids of significant viscosity, viz. $v < Q^{1/2} \cdot H^{1/4}$, first find the equivalent Q–H duty for a liquid viscosity of 1 cSt before entering Fig. 25.2. Correct the hydraulic efficiency thus found by using Fig. 24.8. A further cor-

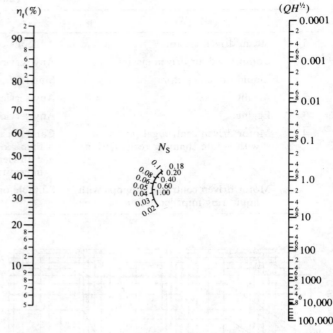

Fig. 25.2. Nomograph for estimating pump hydraulic efficiency at the maximum efficiency point for single stage pumps

η_1 = pump hydraulic efficiency (per cent)
Q = volume flow through pump (l/sec)
H = differential head (per stage (m))
N_s = specific speed (non-dimensional) = $(N \cdot Q^{1/2})/(175 \cdot H^{3/4})$ where the constant 175 includes g, taken as the value 9.81 m/s²
N = impeller rotational speed (r/sec)

Notes
(1) Q is the discharge flow and is not altered for double entry impellers
(2) The efficiency prediction is for a liquid viscosity of 1 cSt
(3) The data upon which this nomograph is based was obtained chiefly from pumps running at 50 and 25 r/sec

rection is required for slurries by multiplying by the factor obtained from Fig. 24.10.

(b) For commercial applications, the pump efficiency is given by

$$\eta_1\left(0.85 + \frac{\eta_1}{1000}\right)$$

where
η_1 = estimated percentage pump hydraulic efficiency for a single stage at the best efficiency point

Because of their internal construction, slurry pumps are intrinsically less efficient. Apply the following additional factors:

light slurry pumps   0.90
heavy slurry pumps   0.80

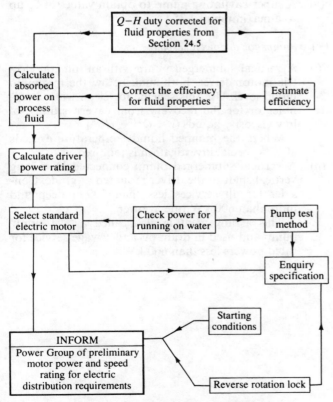

Fig. 25.1. Driver power rating

86

(c) Multistage pumps suffer additional losses due to the guide vanes redirecting the liquid to the eye of the next impeller. Enter the nomograph (Fig. 25.2) using the head per stage and multiply the estimated efficiency by 0.94.

As a first estimate for pumps running at 50 r/sec take the number of stages as

$$Z_0 \sim \frac{H}{7 \cdot Q^{2/3}}$$

Where sufficient NPSH is available the speed can be raised to N r/sec and the number of stages reduced to Z where

$$Z = \left(\frac{50}{N}\right)Z_0$$

Limiting parameters for horizontal shaft multistage pumps are:

(i) number of stages $< 620/N$
(ii) maximum head/stage < 660 m
(iii) specific speed for one stage: 0.03 to 0.08

Vertical borehole pumps are restricted in diameter and consequently are normally cone flow type with $N_s \sim 0.16$ and up to 40 stages.

25.2 Calculation of absorbed power

$$E = \frac{10 \cdot Q \cdot H \cdot \rho}{\eta}$$

alternatively

$$E = \frac{0.981 \cdot M \cdot H}{\eta}$$

where

E = absorbed power (kW)

Q = volume flow through pump (l/sec)

H = differential head across pump (m)

ρ = maximum density of liquid (kg/l)

η = pump efficiency (per cent)

M = mass flow through pump (kg/sec)

25.3 Driver power rating

(a) Calculate the minimum power rating of the driver as

$$\text{Power rating}_{min} = \left(1.12 + \frac{0.1}{H^{0.25}} + \frac{1}{\eta}\right)E$$

where

E = estimated power for the duty point (kW)

H = total differential head (m)

η = pump efficiency (per cent)

(b) Where a speed changer is inserted between the pump and driver, multiply by the appropriate factor given below

Gearbox transmitting E kW	$1.04 + \dfrac{2}{E}$
Belt drive (<220 kW)	1.07

(c) For an electric motor drive select the next highest power rating from the table in section 25.4.

(d) Check the service considerations:

(i) Uprating:
pumps with integral motors should have at least a 10 per cent power margin so that a larger impeller can be fitted.

(ii) Emergency service:
certain duties require that there should be no risk of trip due to motor overload and the motor should be rated accordingly.

(iii) Duties where the process liquid is significantly less dense than water:
recalculate the absorbed power for a density of 1.0 kg/l. If the power rating of the motor is less than this value and less than 100 kW, then adjust the motor rated power to the next size. This permits performance testing on water. Larger pumps require individual consideration.
Process reasons may dictate that the pump be capable of running on water.

25.4 Preliminary power ratings of electric motors

Use one of the following power ratings in kW irrespective of speed

1.1	1.5	2.2	3.0	4.0	5.5	7.5	11	15	18.5
22	30	37	45	55	75	90	110	132	150
185	220	250	275	300	350	400	450	500	550

Consult a specialist to determine the appropriate motor rating and type for variable speed drives.

Notes
(1) This power rating is used only for a *preliminary* estimate of the electrical distribution load.
(2) The frame size and power rating of the motor is decided after selection of the pump, using the pump manufacturer's data.

25.5 Starting conditions for electric motors

When specifying a glandless canned motor pump note that the switchgear may be required to take abnormally large current at starting ($8 \times$ FLC). Confirm this factor after selection of pump supplier.

If a reverse rotation lock has been specified, consider the requirements arising from incorrect connection which subjects motor and switchgear to full voltage locked rotor conditions.

Pumpsets fitted with flywheels for alleviating pressure surge effects may require special motor starting arrangements because of the long run-up times.

Deepwell pumps require a time delay on restart to permit the liquid column to discharge through the pump. Check that this is included in the motor permissive start system and is shown on the line diagram.

25.6 Reverse flow and reverse rotation

Check the line diagram to ensure that a non-return valve is provided on the discharge of each pump when:

(a) two or more pumps run continuously in parallel;
(b) automatic or remote start is required so that a standby pump is left at immediate readiness with open isolation valves;
(c) it is necessary to prevent reverse flow for process reasons when a pump trips out;
(d) overpressure protection of the pump inlet is required – see section 26.5(b).

Specify a reverse rotation lock for a deepwell line shaft pump.

If the delivery vessel is not under gas pressure the reverse rotation speed is unlikely to exceed 125 per cent of the forward speed. The preferred arrangement is that the driver can withstand this reverse speed.

If the pump delivers into a vessel against gas pressure, then upon a pump trip the liquid content of the delivery line will discharge back through the pump, followed by gas discharge. Process reasons may dictate prevention of such gas discharge by the installation of high-integrity non-return valve(s). If this is not done then consider a reverse rotation lock to avoid pump and driver overspeed upon gas discharge. As a rough guide it is worthwhile investigating this in conjunction with the chosen supplier when

$$\frac{\eta_1 \cdot \rho}{\rho_g} > 12$$

where

ρ_g = true gas density in delivery vessel at rated conditions (kg/m^3)
η_1 = pump hydraulic efficiency at BEP (per cent)
ρ = rated liquid density (kg/l)

CHAPTER 26. CASING PRESSURE RATING

The process for the determination of casing pressure rating is modelled in Fig. 26.1.

26.1 Calculations of maximum inlet pressure

$$\text{Maximum inlet pressure} = P_{im} + \frac{\rho_m \cdot h_{sim}}{10.2}$$

where

P_{im} = maximum gas pressure in the inlet vessel, taken as the relief valve pressure setting (bar g)

Note that in closed systems exemplified by refrigeration plant, the maximum gas pressure is usually determined by the 'settle-out' pressure of the system upon trip of the associated gas compressor

ρ_m = maximum liquid density (kg/l)

h_{sim} = maximum static liquid head over the pump inlet, taken to the overflow point, assuming failure of the primary level control system (m)

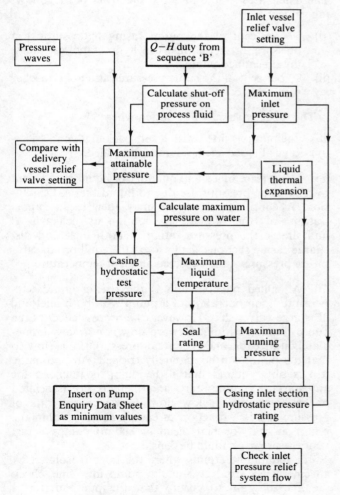

Fig. 26.1.   Casing pressure rating

Notes

(a) The pressure loss in the inlet line is ignored because the pump is assumed to run against a closed discharge valve.

(b) For some pumps the inlet connection of the casing is not rated for discharge pressure. For rating purposes the maximum attainable pressure should be taken as either the sum of the maximum pressure from above and the wave pressure from section 26.3 or, alternatively, as the thermal equilibrium pressure from section 26.4, whichever is greater.

26.2   Calculation of differential pressure at shut-off

(*a*) As a first estimate take the shut-off head as 110 per cent of the differential head at normal flow, alternatively 115 per cent of the differential head at maximum flow, whichever be higher. Where the pump has been specified to have a stable characteristic take the shut-off head as 125 per cent of the rated differential head at normal flow. Redetermine this value when the actual pump characteristic is available.

Add a further 10 per cent to allow for:

tolerance permitted on design and measurement of pump performance;

variations encountered in pump manufacture of nominally identical spare components.

Calculate the differential pressure from the shut-off head using the *maximum* density of the liquid.

(*b*)   The maximum density of the process liquid may be less than 1.0 kg/l. Tests at the manufacturer's works are usually performed on water with atmospheric inlet pressure. For such tests check that the pressure at discharge when operating at the rated $Q-H$ duty is less than the casing pressure rating; if it is greater then special test arrangements will be required.

26.3   Pressure waves

For pumps delivering into long piping systems, the rating of the casings should take account of the additional pressure due to deceleration of the liquid.

For centrifugal pumps the probability of system resonance occurring is small and will be ignored.

(*a*)   *Fast changes in flow*

Fast changes in flow are arbitrarily defined by

$$t < \frac{3L}{C}$$

where

L = total length of the inlet and delivery pipework (m)

C = velocity of sound in the liquid (m/sec)

t = time for the valve controlling the flow to travel from full open to shut (sec)

Such fast changes arise from:

(i)　emergency flow shut-off by a solenoid trip valve;
(ii)　air failure on the normal diaphragm motor for the control valve;
(iii)　manually operated cocks or similar quick-acting valves.

Check the line diagram to see that these causes of fast shut-off are located near the delivery end of the line; if any are located at the pump end then a more comprehensive final analysis should be made.

Check that non-return valves are specified as the damped action type.

Note that power failure or trip out of the driving electric motor initiates a pressure wave. Take the maximum pressure rise as 50 per cent of the differential pressure across the pump.

Some large systems merit reduction of this surge by the addition of a flywheel to the pumpset. In petro-chemical plant this device may be encountered in cooling water systems.

For such fast changes of flow in pipelines, the head rise is given by

$$H_w = K \cdot V \cdot C$$

where

H_w = incremental head rise due to wave (m)

C = velocity of sound in liquid at operating temperature and pressure (m/sec)

Note that both occluded gas and gas slugs markedly decreased the effective velocity of sound in the liquid but the latter do not affect the head rise calculated by ignoring their presence.

K = a factor taken as 0.09 for typical steel pipes

V = velocity of liquid in delivery pipeline at maximum flow (m/sec)

If the piping varies in diameter take the equivalent velocity as

$$\frac{\sum (v \cdot l)}{L}$$

where

v = velocity within a section of length l of the total line length L

(*b*)　*Slow changes in flow*

When the flow changes are slow, i.e.

$$t > \frac{3L}{C}$$

the effects of compressibility can be ignored and the incremental acceleration head calculated as

$$h = \frac{\sum (v \cdot l)}{10t} \quad \text{(m)}$$

where

l = length of line section in which the liquid velocity is v m/s (m)

Where a steady flow is essential, the acceleration head due to operation of automatic control should be limited.

(*c*)　*Lever operated cocks*

The use of lever operated cocks as pump inlet isolation valves can be dangerous: check that wheel and screw valves are used to preclude rapid operation.

26.4　Pressure due to liquid thermal expansion

(*a*)　If a pump handling a liquefied gas at sub-ambient temperatures can be completely isolated whilst full of liquid the heat-leak from atmosphere will raise the pressure in the pump.

(*b*)　Check the line diagram to ensure that a relief valve is fitted on the pump side of the inlet isolation valve. The relief valve should be set at the lower of the following values:

(i)　67 per cent of the pump casing hydrostatic test pressure, or of the inlet can hydrostatic test pressure on canned pumps;
(ii)　90 per cent of the static pressure rating of a mechanical seal.

The relief valve may be omitted when:

(iii)　the pump is fitted with a soft-packed gland;
(iv)　a gas-loaded accumulator is provided.

(*c*)　The release of a small quantity of liquid is sufficient to relieve the pressure rise due to liquid thermal expansion. A continuing heat-leak may then begin vaporization of the liquid. If continuous leakage is undesirable the pressure rating of both inlet and discharge casing sections shall exceed the equilibrium saturation pressure at the expected ambient temperature.

(*d*)　An allied phenomenon is the rise in liquid temperature upon operation of a pump with both inlet and discharge valves shut. However, the process effect of the pump running without forward flow is normally immediately noticeable and the rate of pressure rise is slow so that the pump can be manually tripped. An automatic trip is not required unless the pump is intended for autostart or remote start, and there is no associated process warning for low flow. However, the rate of pressure rise is a function of energy density within the pump so that for high head (>800 m) pumps a fast temperature trip should be considered.

For flammable liquids where the tank is isolated by an emergency trip valve in the pump inlet line, check the line diagram to ensure that the pump driver is simultaneously tripped.

26.5 Casing hydrostatic test pressure

(a) Casing discharge section

(i) In conventional arrangements the pump may run against a suddenly closed valve in the discharge line. The maximum attainable pressure is then the sum of:

   maximum inlet pressure;
   maximum differential pressure;
   wave pressure.

   As a first estimate take the hydrostatic test pressure as 150 per cent of this maximum attainable pressure or 150 per cent of the system pressure relief valve setting, whichever is the greater.

(ii) Hot pumps (handling liquid above 175°C) require individual consideration taking into account piping loads and the reduction in material strength with increase in temperature.

(iii) Small pumps to BS4082 and BS5257 have standardized casing test pressures as follows

BS4082	(Class R)	60 bar g
	(Class L)	21 bar g
BS5257		24 bar g

(b) Inlet connection rating

The rating of the inlet connection should be identical to that of the discharge connection, except for the following.

(i) Vertical canned pumps, where the inlet can may be rated for the maximum attainable inlet pressure.

(ii) Multistage pumps where the discharge connection is rated at more than 70 bar g whilst the maximum attainable inlet pressure is less than 25 bar g.

(c) System arrangements

For these cases at (b) above check the line diagram to ensure that:

(i) two non-return valves in series are provided at each pump discharge.

(ii) a pressure relief valve is provided, located between the pump inlet and the inlet isolation valve, rated to pass at least the largest expected leakage flow-rate past the non-return valves.

PART SIX

Centrifugal pumps: special purpose multistage

CHAPTER 27. SPECIAL PURPOSE MULTISTAGE CENTRIFUGAL PUMPS

27.1 Introduction

It is not the intention of this publication to cover the requirements for the specification of boiler feed pumps for large power station duties. However, many process plants generate steam for internal use, either on direct fired boilers or in waste-heat boilers, over a wide range of pressures. In addition, other high-head duties include injection of boiler-feed water for steam temperature control, main oil line pumping, and injection of liquid into gas scrubbers or reactors operating at high pressures. The following sections čover the additional requirements (over those laid down in Part Five) that need to be considered when dealing with high-head multistage centrifugal pump duties.

27.2 Choice of number of pumps

High-head duties most frequently arise from a need to inject liquid into gas scrubbers or reactors operating at high gas pressures. Process requirements for exit gas purity or product quality often demand 100 per cent availability. Then a standby pump on autostart for immediate readiness is required.

Arrangement 1

One 100 per cent duty pump together with one standby pump. Common practice is to make the standby pump identical to the main pump on grounds of:

(a) common spares holding;
(b) common maintenance and operational methods;
(c) no process interruption upon changeover of running and standby pumps;
(d) identical piping and layout arrangement, minimizing design effort.

Arrangement 2

Two running 50 per cent duty pumps with one identical standby pump. This arrangement may be selected in order to:

(a) reduce the NPSH requirement to avoid the need for inlet booster pumps;
(b) ensure continuity of liquid flow (This condition is important when the delivery system includes heat exchangers which can quickly generate vapour upon flow stoppage.);
(c) increase overall power efficiency when the turndown is expected to be more than 50 per cent of plant rating for more than 50 per cent of the running time by allowing operation with one pump;
(d) relax constraints on steam system control when steam turbine drivers are used;
(e) meet constraints imposed by commercial availability of advanced pumps.

27.3 Review of alternative pump types

Barske type

The high-speed one or two stage Barske type pump requires a high NPSH and may have an unstable Q–H characteristic. Nevertheless, where the process demand varies widely, with campaigns of low flow operation, consider the use of three or more such pumps in parallel so that the number of running pumps can be adjusted to meet the demand.

Peripheral type

The peripheral type pump is well suited to small capacities (<2 l/s) because its Q–H characteristic is inherently stable. However, it is essential that the process liquid is sufficiently pure and free from suspended solids to make acceptable this pump's sensitivity to erosion. Specify this type of pump in preference to reciprocating pumps for heads up to 600 m, where the liquid is clean.

27.4 Duty

This is calculated using Part Five amended as follows.

(a) Differential head

Calculate the required differential head at maximum flow, with throttle control valves *wide-open*, and with the delivery vessel pressure taken as the set pressure of the vessel pressure relief valves.

A correction to the simple head calculation may be required; refer to Appendix VI.

(b) NPSH

When calculating available NPSH note that most vessel reservoirs have only single element level control; this demands either a generous NPSH allowance for acceleration head or a heavily damped throttle control valve, to cope with transient upsets.

High-speed large capacity pumps may require unusually large values of NPSH. As speed or capacity increases, keeping S_n constant maintains hydraulic performance but the intensity of local cavitation increases. There is a limiting value of NPSH for a given material and impeller construction. For typical castings in 13 per cent Cr steels take this limit as defined by

$$\text{NPSH} > \frac{N^2 \cdot Q}{44\,100}$$

where the values are taken at the pump best efficiency point and apply over the operating range 85–105 per cent of the capacity at BEP; the S_n should be further reduced if a wider range of operational capacity is required.

Sudden failure of the steam or hot vapour supply to gas strippers gives a transient reduction in the available NPSH. If the residence time for the stripper vessel reservoir is less than 600 seconds, a special investigation is necessary.

For the available NPSH to be a maximum during such a transient the inlet line should be sized to give a liquid velocity at rated flow of 2–5 m/s, the higher value applying to sensitive systems with short residence times.

(c) Capacity

The flow must be continuously above a critical minimum value set either by NPSH considerations or by system control stability requirements.

Typical pumps have $N_s \sim 0.06$ for which the minimum flow is greater than 20 per cent of the pump capacity at BEP. Small high-speed Barske pumps require a minimum flow of the order of 60 per cent of the BEP flow.

The highest system reliability is obtained only when the minimum flow is provided by a simple continuous bypass through a let-down multiple restrictor back to the inlet vessel. The capacity of the pumps is then increased to supply both the bypass and the desired delivery flow.

Large pumps may merit a dedicated control system to maintain minimum flow by opening the bypass valve when the delivery flow falls below the specified minimum flow. Such a system should be of high integrity.

27.5 Estimating the number of stages

For horizontal shaft multistage centrifugal pumps, the choice of the number of stages is not a precise determination. Some guidance is given in section 25.1(c) and the duty field in Fig. 25.12 of Part Five.

For small pumps where $Q \cdot H^{1/2} < 200$ or where the efficiency is not important

$$\frac{2 \cdot H}{Q^2} < Z > \frac{8 \cdot H}{N^{4/3} \cdot Q^{2/3}}$$

As the number of stages increases the bearing span increases and the rotor becomes more sensitive to dynamic effects. Multistage centrifugal pumps usually run at operating speeds above the first lateral critical speed obtained by assuming the pump to run on air; they rely on the damping provided by the pumped liquid to suppress dynamic effects. Unwanted perturbations arise from operating at part-capacity, producing large low-frequency radial hydraulic forces.

The pump shaft diameter and bearing span may be varied by the pump designer for a given hydraulic duty, rotor speed, and number of stages. Guidance on acceptable relationships of these parameters is given in Fig. 27.1 which is based upon avoidance of excessive shaft bush wear. Note that such wear has led to shaft fracture in Zone A.

Values of d and L are not available, unless the uprating of an existing pump is being investigated. As a first estimate take the limiting number of stages as

$$Z \sim k\left(\frac{Q^2}{N \cdot H}\right)^{1/3} - 2$$

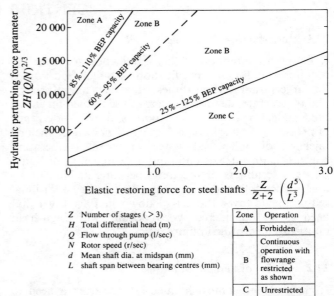

Zone	Operation
A	Forbidden
B	Continuous operation with flowrange restricted as shown
C	Unrestricted

Z   Number of stages (> 3)
H   Total differential head (m)
Q   Flow through pump (l/sec)
N   Rotor speed (r/sec)
d   Mean shaft dia. at midspan (mm)
L   shaft span between bearing centres (mm)

Fig. 27.1. Influence of shaft stiffness on choice of number of stages for horizontal multistage pumps with radial face mechanical seals

where

$k = 11$ for stiff rotors, capable of momentary 'dry' running

and

$k = 51$ for conventional rotors operating continuously within the flow range 85–110 per cent of BEP capacity.

A double-entry first stage is the equivalent of two single-entry stages for this calculation.

The number of stages is reduced by increasing the pump rotational speed but this increases the NPSH required by the pump.

Inlet booster pumps are needed when the NPSH available is insufficient to meet the NPSH required by the main pump. A common arrangement has the booster directly coupled to a double ended electric motor which also drives the geared main pump. The booster may itself be a multistage pump to reduce the duty of the main pump.

Where the limiting number of stages is less than the number of stages required to obtain a reasonable efficiency, then consider a vertical shaft pump.

27.6 Variable speed control

The chief considerations are:

(a) Commercial speed governors for small steam turbines have deadband zones of about 0.25 per cent when specified to NEMA Class 'D'. For typical pump and system characteristics, this limits the stable range of control on a single pump to 120 to 60 per cent of BEP capacity; below this range ordinary throttle valve control is used with the speed held constant.

Arrangements where two pumps run in parallel

exacerbate the stability problem. A successful method is to run the turbines at constant speed and rely on throttle regulation, with a control valve in each individual pump discharge line.

(b) Fluid couplings can provide significant excitation torques at discrete frequencies and have been identified in one case as the cause of impeller failure. Variable speed electric motors of the commutator type can similarly provide excitation torques.

27.7 Drivers

(a) Steam turbines

The advantages of using different designs of main and standby steam turbines are sufficiently large to outweigh the reasons given in section 27.2 for selecting identical main and standby pumps.

Cost considerations favour the arrangement where the main pump is driven by an electric motor, whilst the standby pump driver is a turbine appropriate to slowroll and quick-start duties.

Unfortunately, the conventional small turbine for these latter duties normally has an efficiency of the order of 30 per cent. Such a turbine imposes a sudden high demand on the steam supply system upon starting. In process plants these turbines are normally supplied from an intermediate pressure steam header. Conventional control systems cope much better with reductions than with increases in steam demand; consequently for sensitive systems the preferred driver for the standby pump is an electric motor, whilst the main pump is driven by a high-efficiency steam turbine.

Standby turbines of conventional construction may be held on slowroll to ensure instant readiness. For this purpose, current practice is to leave the emergency steam stop valve open, and the inlet autostart valve closed but bypassed by a small restrictor whose orifice size is empirically adjusted to give the required slowroll speed. The autostart valve actuator is damped to give a valve stroking time of 10 seconds (for a linear valve characteristic) in order to avoid any pump speed overshoot.

For the emergency condition of electrical supply failure, consider whether the turbine driver need be capable of developing rated power when exhausting either to a secure steam system or to atmosphere via a pressure relief valve.

(b) Electric motors

It is expensive to obtain slowroll operation on an electric motor driven pumpset. For standby duty, first consider the set as stationary and review the consequent effects on both pump design and layout.

Standby pumps are preferably direct-drive. Gears that remain stationary need special consideration.

(c) Dual drivers

One pump may be coupled to two drivers, each capable of driving the whole set. Such an arrangement gives a cost saving for large pumps when the reliability of the normal driver is much lower than that of the pump. This case arises when the normal driver is an electric motor connected to an unreliable electric supply system and the alternative driver is a steam turbine. Now the failure rate of pumps is of the order of 0.5/year, consequently this arrangement is not justifiable for UK sites with access to the CEGB grid, when the supply failure rate is of the order of 0.1/year.

The dual drive arrangement may be used for power recovery, where a steam or gas turbine is in continuous operation and the motor can act as an induction generator exporting surplus energy as electrical power.

27.8 Installation requirements

(a) Transient effects

Large vessels containing vapour, or process heat exchangers, in the pump delivery system constitute accumulators capable of supplying vapour which may run the pumps in reverse rotation. Current practice is to provide reliable non-return valves in each pump discharge line rather than reverse rotation locks.

(b) Temperature gradients in casings of hot pumps

Non-return valves leak a small amount. Such leakage can lead to thermal stratification in the casing and thence to casing distortion. Preventive measures are:

(i) The elimination of temperature gradients by slowrolling the standby pump.

(ii) The provision of a bypass round the discharge non-return valve in order to ensure a sufficiently large reverse flow of hot liquid. The required flowrate depends on pump size and layout but is typically 2 per cent of the pump rated capacity.

It is *essential* to maintain adequate thermal insulation around the pump casing.

A simple hole drilled through the non-return valve element suffers erosion and generates noise: the recommended arrangement is an external piped bypass embodying a multiple restrictor which can be empirically adjusted during commissioning.

This arrangement re-introduces the risk of reverse vapour flow from process heat exchangers; consequently a reverse rotation detector should be considered to trip shut the discharge block valve.

The distribution of flow through the stationary pump is indeterminate unless the natural downward drift of cooled liquid is encouraged. The detail arrangement of the pump may result in stagnant zones. Extraction of cool liquid from such zones will require the provision of an ancillary pump for return to the inlet vessel or disposal elsewhere.

(iii) Temperature gradients can be prevented by positively isolating the pump to create a truly stagnant liquid condition. The pump then has to withstand 'cold start' conditions of thermal shock. Both these requirements are difficult to realise; consequently this scheme needs careful engineering.

PART SEVEN

Sealing considerations

CHAPTER 28. SEALING CONSIDERATIONS FOR ROTATING SHAFTS

Part Seven covers the requirements for the specification of shaft sealing for both rotary positive displacement pumps and centrifugal pumps.

The problems of sealing rotating shafts are quite complex and require careful consideration. In the last analysis the sealing problem can affect the choice of pump selected to handle the duty.

The system for dealing with sealing considerations is modelled in Fig. 28.1.

28.1 Preliminary choice of seal

(a) Refer to any previous operating experience or to seal manufacturers for practice appropriate to specific liquids.

(b) Specify soft-packed glands for:

(i) water pumps in emergency or fire-fighting service;

(ii) drain pit pumps or similar water service duties where intermittent wet/dry running is expected.

(c) Make glandless pumps the first choice for:

(i) duties where the liquid is toxic or obnoxious so that no leakage can be permitted;

(ii) pumps in biochemical service where sterility is essential;

(iii) pumps handling liquids where the inlet pressure is above 43 bar g.

For rotary pumps, except for small pumps in Family 1 (see section 17.6(a) and Table 17.1),

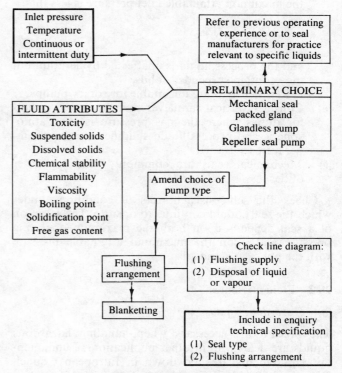

Fig. 28.1. Sealing considerations

glandless construction is not commercially available.

See Table 28.1 for type classification of glandless pumps.

Table 28.1. Glandless pumps – type classification

Family	Member	Power (kW)	Application notes
Self-contained	Canned magnetic coupling	120	(1) Where separate driver needed, e.g., steam turbine (2) Process liquid temperature > 110°C and cooling is inadmissible
	Canned electric motor with dry stator	150	External ancillary cooling system required if process liquid temperature > 110°C and such cooling permissible
	Submersed wet stator electric motor	2500	Water duties where water temperature in motor can be limited to 40°C, using ancillary external systems. Other duties possible if process fluid compatible with insulation
	Hydraulic turbine/pump units	—	Booster and 'down-hole' duties
Buffer System	Self-contained family with auxiliary liquid injection system	—	Auxiliary supply provided by positive-displacement pump to guarantee flow insensitive to pressure variations
	Submersed water-filled electric motor separated from pump unit by mechanical seal	60	Intermittent use
	Gas pressurised vertical shaft immersed pump	—	Liquids that solidify below 100°C

(d) Consider repeller seals for unspared pumps where the maximum attainable inlet pressure is less than

$$\frac{\rho \cdot N^2}{1000} \quad \text{(bar g)}$$

and the process liquid is unaffected by aeration but may contain suspended solids.

Repeller seals are unsuitable for rotary pumps.

(e) Consider vertical cantilever shaft immersed pumps for duties where the gas pressure is near atmospheric and process liquid solidifies at a high temperature ($> 100°C$).

(f) Mechanical seals are commonly used for other duties.

Check the self-consistent sets of conditions under which the seal should continue to operate. The selection of a seal depends upon both the maximum operating pressure and upon the maximum attainable pressure with the pump at standstill.

28.2 Fluid attributes

(a) Hazardous liquids

The precautions necessary when handling hazardous liquids are dealt with in other publications. Fundamental requirements are laid down in European Council Directive 82/501/EEC (Major Hazards of Certain Industrial Activities), applied in the United Kingdom by Statutory Instrument 1902 (Control of Major Accident Hazards: 1984).

At this initial stage of process pump selection the crucial choice lies between glanded and glandless pump constructions.

All glands leak. A practical minimum leakage is obtained by using a well-engineered rotary radial-face mechanical seal, when a representative value of 10 ml/h can be taken for a typical pump in petrochemical plant service. Such a seal may appear to run dry because the leakage emerges as vapour.

The vapours from some process liquids are hazardous only in relatively high concentrations. Then, the small continuous leakage from glanded pumps becomes a problem only when the pumps are installed in a confined space.

Confined spaces are plant locations that are poorly ventilated, i.e., there is little or no air movement to disperse the vapours or mists resulting from leakage. Alternatively they are spaces where there is no quick means of escape or quick remedial action available.

Examples of these areas are:

(i) inside a building that is substantially closed;
(ii) outside locations that are surrounded by walls, buildings, or other equipment forming sheltered pockets where vapour can accumulate;
(iii) pits deeper than 1 metre and of a breadth less than five times the depth, which can retain dense vapours, particularly cold ones arising from low-boiling-point process liquids.

Questions to be considered are:

(iv) is the ventilation rate adequate to disperse vapours or fumes quickly? (Forced ventilation systems must have an adequate level of reliability.);
(v) are personnel close to the pumps for long periods?
(vi) are there at least two exits in different directions for means of escape?

Pools of liquid can accumulate from the small leakage of stable, high-boiling-point process liquids; alternatively they can arise from gland failure. Their disposal is part of the larger problem of the safe disposal of process liquids and vapours from the whole plant.

Where neither vapour nor liquid leakage can be accepted, the self-contained glandless pump is the first choice (see Table 28.1). Where the duty is beyond the limits imposed by such pump constructions a special investigation is necessary.

(b) Solid particles suspended in the liquid

Specify appropriate seal design and arrangement by reference to works experience on similar installations.

For nominally clean liquids, check the following possible sources of particulate contamination:

(i) ceramic particles formed by attrition in towers packed with ceramic rings;
(ii) particles remaining in the mother liquor after a primary filtration or centrifuging process;
(iii) particles accumulated in the circulating liquid of a gas scrubber when the blow down is small and the gas is likely to contain solid particles, e.g., catalyst dust;
(iv) overload in circulating systems for dissolving tanks used to recover a solid product;
(v) particles captured during condensation on cooling a process gas, forming a contaminated drain liquid;
(vi) occasional operation, giving sufficient time for corrosion products to form in vessels and lines;
(vii) commissioning runs of systems that cannot be cleaned.

Liquids can be regarded as clean when the pump draws from a tranquil storage vessel large enough to permit settling of the solid particles, provided that the outlet is not at the bottom.

(c) Solids in solution

(i) *Crystallisation on cooling*
Hot concentrated solutions are prone to deposit crystals both inside and outside the seal. Consider an injection flush of the solvent, together with a quench.

(ii) *Crystallisation on evaporation*
Leakage of cool but concentrated solution evaporates leaving the solids deposited outside the seal face. Specify a quench flush.

(iii) *Waxing*
Some liquids may contain waxy polymers. Specify provision for injecting a solvent into the sealbox at pump shutdown, and consider whether heating is necessary.

(d) *Chemical and physical properties*

(i) *Chemical stability*
If the liquid decomposes with increase of temperature or decrease of pressure to form solid products, then specify an injection flush. An example of this is copper liquor (cuprous ammonium acetate) which deposits metallic copper between the seal faces.

If the liquid is oxidised on contact with atmosphere to precipitate a solid product then specify a quench flush.

(ii) *Viscosity*
Liquids of high viscosity can impose heavy loads on the seal driving mechanism. Check the line diagram for heating arrangements for inlet pipes, vessels, and pump stuffing boxes, particularly for standby pumps.

(iii) *Liquids near their boiling points*
The pressure in the seal box should exceed the vapour pressure by an increment corresponding to a 15°C temperature rise.

Check that the liquid can be cooled to its boiling point at atmospheric pressure. If it cannot, a special investigation is needed.

(iv) *Liquids near their solidification points*
In mixtures having no well-defined freezing point, the viscosity reaches very high values. For these and for hydrocarbon liquids take the effective freezing point as the 'pour-point' temperature.

The seal and pump should be jacketed and heated by steam or by hot liquid from another source. Note that the steam pressure is fixed by the temperature required: check the line diagram to ensure that a suitable supply is shown.

Where the solidification temperature exceeds 100°C consider use of a glandless vertical cantilever shaft immersed pump.

(v) *Free gas content*
Experience indicates a marked reduction in mechanical seal life when the liquid contains occluded gas. Specify injection flush or double seals with circulation flush.

Table 28.2. Definition of flushing arrangements

Arrangement
Process liquid circulation flush
Some process liquid is taken from a connection on the pump casing, passes over the seal, and returns through the seal box to the pump inlet.
Injection flush
The flushing liquid is obtained from an external high pressure supply and injected into the seal box. It passes over the seal, then through an inner restrictor in the seal box to join the main pump flow.
Where the inner restrictor is a reversed lip seal, take the minimum injection flush rate as 2 l/hr/mm of shaft sleeve diameter.
The injection flush mixes with the main pump flow and should be both compatible and miscible with the process liquid over the operating temperature range.
Barrier liquid system
The barrier liquid is obtained from an external supply and is circulated between double seals arranged back-to-back.
The pressure of the circulating flush liquid should always be higher than the pressure in the pump seal box at the inner seal by at least 2 bar.
The pump main flow is contaminated by leakage from the inner seal. The flushing liquid should be compatible with the process liquid.
Quench flush
The quench fluid flows through the chamber between the main seal and an outer auxiliary seal arranged in tandem.
The pressure of the quenching flush should always be *lower* than the pressure in the pump at the main seal.
Steam is often used as the quench fluid.
For vacuum systems where air ingress is hazardous, specify a quench flush using either a liquid or an inert gas for blanketting on shut-down.
Zero flush
Sometimes called 'dead-end'.

(vi) *Liquids at sub-zero temperature*
Pumping temperatures below −3°C cause all external surfaces to ice up by atmospheric deposition. This hinders free axial movement of the flexible member. Specify an external auxiliary gland to enclose the main seal in cold dry leakage vapour.

ADDENDUM

Pump and system combined

(Extract from EEUA Handbook No. 30 *A Guide to
the Selection of Rotodynamic Pumps*

PUMP AND SYSTEM COMBINED

(a) Duty points and combined characteristics

When a pump is matched to a system, that system's Q–H characteristic, superimposed on the Q–H characteristic of the pump, will cross at a point satisfying both system and pump characteristics. This is the stable point of quantity and head at which the pump freely operates in the system, and is called the 'duty point'.

It should be noted that the characteristics are the overall curves for both pump and system. In the case of the pump, any H value is the manometric total head developed across the pump. In the case of the system, all losses are included from entry to exit, and all differences of elevation taken account of algebraically. Typical combinations are shown in Fig. Add.1.

(b) Parallel and series operation

When pumps are operated in parallel, or in series, the head on the system and the quantity *are defined* by the point of intersection of the system resistance with the new *combined* pump characteristic. For series operation at any point value of Q, the *head* of that point is increased by the number of pumps in series. For parallel operation, for any point value of H, the *quantity* is increased by the number of pumps in parallel.

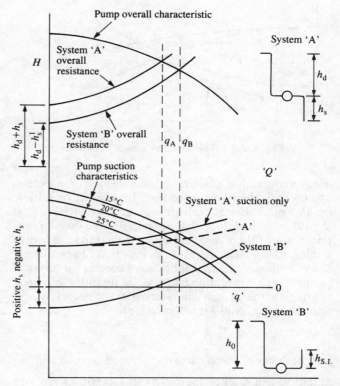

Fig. Add.2.   Effect of changes in suction conditions

It is possible, as shown in Fig. Add.3, for an increase in capacity (or head) to be obtained by either parallel or series working, and a decision on which to employ depends largely on the shape of the system characteristic. With a characteristic (a), a substantial increase in Q can only be obtained by parallel working, whilst for the (b) system resistance, the pumps have to be operated in series. The combined series and parallel curves cross at S, and if the new system characteristic lies to the left, series multiplication is indicated, and vice versa.

Generally, series operation should not be used, unless it is unavoidable, owing to questions of reliability in operation and difficulty of maintenance. If one pump

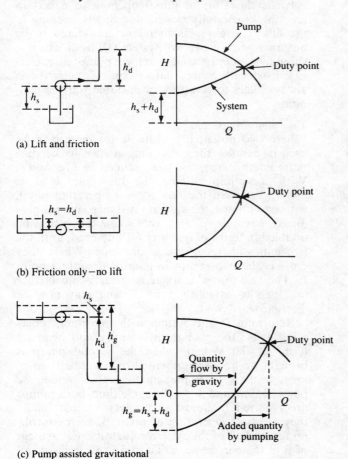

Fig. Add.1.   Typical combined characteristics

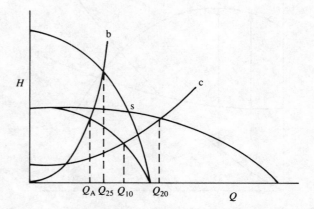

Fig. Add.3.   Choice of series or parallel working

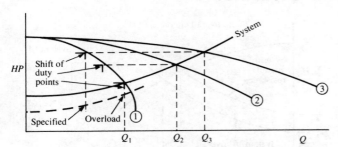

Fig. Add.4.   Three similar pumps in parallel

needs maintaining, elaborate disconnections are necessary to keep the plant operating. In parallel operation, on the other hand, each pump can be valved off. It should be noted too, that in parallel the loss of one pump out of three does not mean a quantitative loss of a third. Conversely, the more pumps that are combined, the less efficient the multiplication becomes, in terms of pump usage, since the increase in quantity obtained relative to the *system* characteristic becomes less and less. From Fig. Add.4 it will be clear that

$$Q_2 < 2Q_1 \text{ and } Q_3 \ll 3Q_1$$

(c)   Stability of pump curves

Stability is in essence the ability of a pump at any one head to produce only one flow condition. Not all centrifugal pumps are so endowed by their characteristic, notable exceptions being those of low specific speed and high impeller vane discharge angle. Such pumps tend to give a humped characteristic curve, of the type shown in Fig. Add.5, from which it will be noted that the 'closed valve' head is less than the maximum head of the characteristic.

All such pumps are potentially unstable, but it should be clearly understood that they may not actually show any sign of instability. Stability is a function of the pump with respect to the system in which it is to operate. The system and pump have to be examined together in two conditions: firstly, considering starting

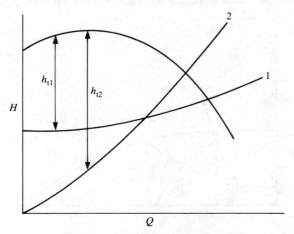

Fig. Add.5.   Unstable pump characteristic

for both single pumps and pumps in parallel; secondly, considering the running condition of the system.

(i)   Starting conditions

(1)   Single pumps

Referring again to Fig. Add.5, the system resistances against the pump characteristic are of two kinds: (1) which is a high static lift, low friction type, and (2), which is a high resistance type. On starting the pump, the duty point of each system is obtained by adjustment of the pump throttle valve, which is eventually fully open (except for any control margins) at the duty point.

Consider now System (2). The excess head to be dissipated in the throttle valve is the vertical distance h_{t2} between the pump curve and the system curve at any point of flow. It will be noted that in System (2), the degree of valve opening (i.e., the length of the throttling ordinate) decreases from 'no flow' to 'full flow', although the pump curve itself is humped. Conversely, with System (1), the distances h_{t1} after the initial valve opening reach a maximum at some point defined by the departure of the two characteristics, and then decrease again to nil at the duty point. Clearly, the action of the valve is then one of initial opening to a certain amount, of carefully closing and finally opening to the full. The pump is therefore considered to be unstable with respect to System (1), from which it should be noted that a humped pump characteristic used with a very flat system characteristic is always likely to result in an unstable starting condition.

(2)   Two pumps in parallel

Where two potentially unstable pumps are to be used in parallel, the conditions of stability become more complicated, and are outlined in Fig. Add.6. With one pump running, the duty point is at A, and to establish the final *system* operating point, with two pumps, the system resistance curve is produced beyond A, and the single characteristic doubled in terms of quantity for any head, until the two-pump characteristic is obtained. Where they cross is the two-pumps-in-parallel operating point B. The two-pump characteristic serves no further use than to establish point B, and may now be disregarded.

Returning to the system with one pump operating at A, the change in quantity and head (in terms of the system) when the second pump is brought on line is represented by a gradual movement from A to B passing through intermediate points 1, 2, 3 and 4. At the same time, both pumps are required to adjust themselves so that finally they are both operating at point B, each contributing one half of the system's quantity requirement of B at the correct head. To make this adjustment, the pump already on line must retreat along its characteristic from A to B′, at the same time as the

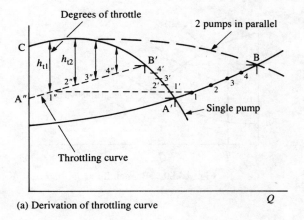

(a) Derivation of throttling curve

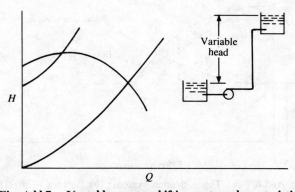

(b) Head dissipated in control valve

Fig. Add.6. Stability of pump starting

pump being started advances along its characteristic from C to B'. In doing this, the first pump is progressively contributing less flow to the system as the second pump progressively contributes more.

Consider now any intermediate point in the process concerned. At 1 the first pump has retreated to point 1' on its characteristic in order to gain the necessary increase of head. A quantity $1 - 1'$ must therefore be obtained from the second pump at this point. Further, it must be delivering this quantity against the head requirement of point 1. Point 1" is therefore established a distance from the origin equal to $1 - 1'$ quantitatively and at the head corresponding to the system resistance of point 1. Similar points are established corresponding to 2, 3, 4 and the interconnecting curve drawn. It will now be appreciated that, in bringing the pump on-line, the excess head to be throttled off in the second pump's control valve is the vertical ordinate above each point on the curve consecutively – at point 1 the throttle is h_{t1}, etc.

To examine the pump for stability at starting, the requirement is now that the ordinates so found should become progressively smaller as the pump approaches B'. When plotted on a separate graph of Q against throttle head, if the curve is progressively diminishing, a stable condition obtains. If this curve is humped, then the pumps are unstable in parallel in the system concerned.

This does not necessarily mean that they cannot be started in parallel, but that such a start is a carefully controlled affair with the characteristic of the control valve as outlined in Fig. Add.6(b).

In the example given, a potentially unstable pump is shown to be quite stable when brought on line in that particular system against a similar pump already working. To bring the first pump on line originally may, of course, prove to be an 'unstable' case, and should be checked independently as previously outlined.

(ii) Operating conditions

Despite the fact that a potentially unstable pump may be brought on line stably, its operation condition may prove in itself to be unstable. To investigate this aspect, consider Fig. Add.7, which shows a system having a variable static lift and two free liquid surfaces. At one condition of lift, the system curve 1 is appropriate and the pump may well be stable. If, however, the demand and head on the system changes, so that system curve 2 applies, it will be apparent that the movement has carried the pump duty point beyond the 'hump' in the characteristic. As soon as the head in the system has increased to the maximum pump head D, then the pump can no longer provide the necessary pressure contribution, and a tendency to instability occurs. The pump tends to return to the no flow condition and, as a result, the resistance of the system drops, consequently causing the pump to increase head and quantity delivered again in trying to arrive at a balance. An oscillation is therefore set up, and the condition of operation becomes unstable.

Such conditions are most likely to occur when the system has a free liquid surface, which may (or may not) be under pressure. It is therefore of considerable importance to be operating in a stable state, with pumps that have fundamentally stable characteristics.

(d)   Maximum power conditions

Where a system is designed specifically for operation in parallel with either one, two, or three pumps in oper-

Fig. Add.7.   Unstable pumps: shifting system characteristic

ation, careful analysis is required to ensure that the pumps are not overloading at the single pump duty point. Figure Add.4 shows three such pumps and the transferred duty points at which single, double, or triple working occurs. If they are specified for the maximum head condition only, the lower head point may cause overloading of the motor, as shown. In such cases, all three duty points should be specified, and the power characteristic should be of the non-overloading type.

(e) 'Effective' pumps and branched systems

The characteristics so far drawn have been those of the complete system measured against the pump *at the pump*. There may, however, be a branch from the system at point A, Fig. Add.8(a), and the effect of the pump on the branch line may be required, when pumping a combined quantity into both legs of the system. It is convenient, therefore, in analysing quantities passing through both system divisions, to imagine the pump as it would be if it were located at A. Consider now the characteristic of the pump plotted against the system's characteristic from D to A, taking h as the elevational difference and h_f the friction losses at any quantity q. The pump's effect in any of the branches is as if the h_f losses had been deducted from the pump's performance, and so the pump's characteristic measured at A will be the reduced dotted line in Fig. Add.8(b). The branch conditions may now be considered on the simple basis that what enters

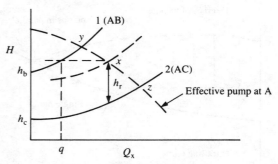

Fig. Add.9.   Branch flows

one branch is not available to the other. In Fig. Add.9, 1 is the system resistance of AB and 2 that of AC. When the duty point of the effective pump is at X, the quantity passing through system 1 will be q_1, and through system 2 will be $(Q - q_1)$. Clearly, at this point, system 2 will be throttled at C to give a residual pressure h_r.

Also, by examination, if point X moves to Y, then Q and q_1 coincide, and there is no flow in system 2 (AC).

Equally, if system 2 is unthrottled, the head (at X) will fall below B and no further flow will occur in AB. Final de-throttling will result in eventual balance on system 2 alone at Z.

Take the case when point Z is above h_b, as indicated in Fig. Add.10.

The two characteristics must now be combined to give the dotted line. The condition for no flow in 2 is the same as before, but 2 can no longer starve 1 entirely of flow. The combined open valve condition is at Z, giving q_1 and q_2, respectively, as the flow in each pipe.

Another example of the use of effective pump characteristic is the insertion of a pressure-reducing valve in the system (Fig. Add.11). Let us assume that system ABCD is protected by a pressure-reducing valve at E. The effective pump at B is required in order to analyse branches BC and BD. Clearly, no pressures of the pump exceeding the pressure-reducing valve setting will be felt beyond E, and the analysis, as before the pump, is superimposed with the cut-in pressure of the valve plus

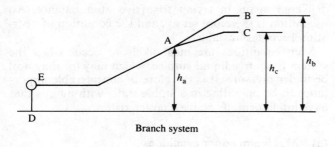

Branch system

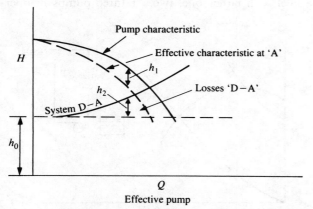

Effective pump

Fig. Add.8.   (a) Top – branch system; (b) Bottom – effective pump

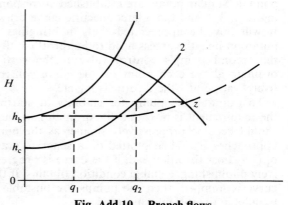

Fig. Add.10.   Branch flows

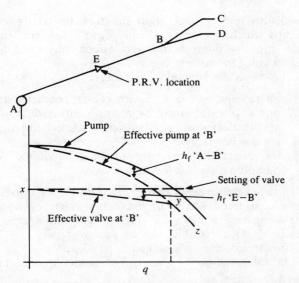

Fig. Add.11. Effective characteristic with pressure reducing valve

the losses from the valve to point B, since the effect of the valve at B is reflected through the pipework EB. Figure Add.11 shows the net effect and it is seen that the pressure-control valve is the effective control for all quantities up to q when the pump controls the system. Flows in branches B and C must therefore be judged against the effective net curve XYZ.

(f) Modifications of duty

Any installed system tends to become modified, as changes in the use of the plant become necessary.

Typical changes of this sort are shown in Fig. Add.12, where the original duty point is A. An increased product demand, coupled with additional pipework,

means a new duty point at B, and a reduced demand on the same pipe system is indicated at C, where a worthwhile power saving may be experienced. In both cases, change in speed or in impeller diameter are indicated.

Changes in speed may well mean changing a motor, and, coupled with the fact that this is usually an expensive modification, precise changes in duty or speed cannot always be met in this way. It is far more likely that such changes to new duty points are solved by changes in impeller diameter: a decrease to meet case C and an increase to meet case B, in Fig. Add.12.

These changes can be calculated

$$\frac{d}{d_1} = \frac{Q}{Q_1} = \sqrt{\left(\frac{H}{H_1}\right)} = \sqrt[3]{\left(\frac{P}{P_1}\right)}$$

but as the impeller is reduced, these rules become approximate due to changes in efficiency and changes in theoretical design assumptions.

The graphical analysis shown in Fig. Add.13 largely eliminates trial and error methods in assessing a revised impeller. The new duty point X is shown in relation to the head–capacity curve of an existing impeller. Any capacity Q_Y is then chosen beyond Q_X and the corresponding head to H_Y is calculated from

$$H_Y = H_X\left(\frac{Q_Y}{Q_X}\right)^2$$

Point Y should lie beyond the known Q–H curve, so that points X and Y can be joined, intersecting the known Q–H curve at Z. The required impeller diameter ratio is then

$$\frac{Q_X}{Q_Z}$$

Similarly, an increase may be calculated. The line XY lies close to the equal-efficiency line through X, and Y and Z have approximately the same specific speed.

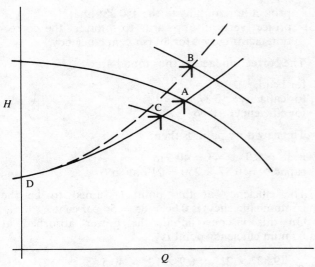

Fig. Add.12. Modifications of duty

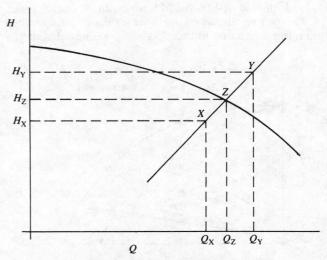

Fig. Add.13. Variations of impeller diameter

The above applies to radial pumps only, and although in principle the same technique can be applied to mixed flow pumps, for constructional reasons not so much cutting of the impeller can be done.

With axial flow pumps, the capacity will vary with the swept area of the impeller, and the head at any point on the impeller varies with the square of the diameter at that point. Reducing the diameter of an axial impeller would require a new ring or new casing, a solution not often used in practice. Variations are obtainable, however, by altering the angle of attack of the propeller blade.

For radial or mixed flow pumps, the question of increasing the impeller diameter at some future date should always be borne in mind when initially buying the pump. One should carefully consider whether any pump should be bought whose frame size will not take a reasonable increase in capacity at a later date. It is always advisable to have 5 per cent at least in hand, to bear increases within any casing, and usually ±15 per cent are acceptable variations, limited on the one hand by the casing size and on the other by loss of efficiency.

Where there is a known possibility of an increase in diameter of impeller, it is also very important for the motor to be capable of dealing with the increased pumping quantity when the impeller is changed. The capacity of the driver must in any case be checked when an impeller is increased in diameter, to ensure that it is not overloaded.

(g) Pumping viscous liquids

When a pump has to move a viscous liquid, the characteristics will be different from those for water. Figure Add.14 shows the characteristics for water in continuous lines and for a viscous liquid in dotted lines.

The greater the viscosity, the higher the power absorbed by the pump, and this may assume such proportions that the use of a centrifugal pump becomes uneconomical. If this is the case, a positive displacement pump (e.g., screw pump) may have to be used.

With the aid of Fig. Add.15, an estimate can be made of the power absorbed by a centrifugal pump when pumping a viscous liquid. It should be noted that the

results are only a rough approximation. Especially with pumps which, on water, already have a low efficiency (low capacity pumps), the calculations may yield too high a value for power absorbed.

The graph can be used in two ways:

(1) for calculating (starting with known characteristics on water) the head/capacity point and power absorbed at maximum efficiency when pumping a viscous liquid;

(2) for calculating the size of pump required for pumping a viscous liquid.

In the example below, using Fig. Add.15, a pump has been considered which, at maximum efficiency point, has a capacity of 250 m³/h (1100 USgal/min) against a head of 43 m (140 ft) when pumping water. The maximum efficiency is assumed to be 85 per cent.

On water, the power absorbed at maximum efficiency point is

$$P = 9.807 \times \frac{250 \times 10^3 \times 43}{60 \times 60 \times 0.85} = 34 \quad \text{kW}$$

To calculate the performance of this pump and the power absorbed when pumping a viscous liquid, the following liquid is considered

specific gravity = 0.9
      viscosity = 440 cSt (2000 SSU)

To arrive at the new duty point, correction factors have to be applied to head and capacity. In addition, a correction factor has to be found for the efficiency. It is evident that the efficiency will be decreased, since capacity and head will be lower and power lost because of the internal friction of the liquid.

Using Fig. Add.15, the example can be developed as follows:

(1) project capacity point, 250 m³/h from the base of the graph vertically upwards to intersect the 43 m line;

(2) project horizontally to the 440 cSt line;

(3) project vertically upwards to intersect the correction factor curves for 80 per cent efficiency.

The correction factors thus found are:

for head: 0.94
for capacity: 0.87
for efficiency: 0.66

The new duty point is then:

head = 0.94 × 43 = 40.5 m
capacity = 0.87 × 250 = 217 m³/h

The efficiency at this point (assumed to be the maximum efficiency) is 0.66 × 85 = 56 per cent.

On this viscous liquid, the power absorbed at maximum efficiency point is

$$P = \frac{9.807 \times 217 \times 0.9 \times 10^3 \times 40.5}{60 \times 60 \times 0.56} = 38.5 \quad \text{kW}$$

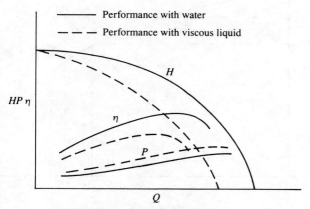

Fig. Add.14.   Viscous liquids

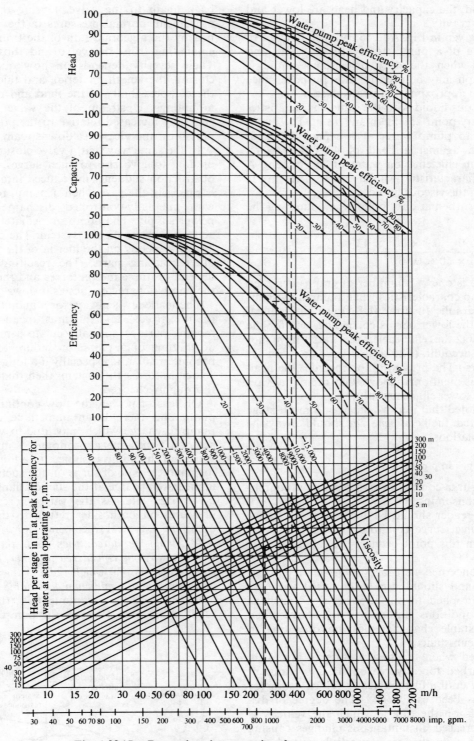

Fig. Add.15. Pump viscosity correction chart
(Dotted characteristics give correction factors
according to the Hydraulic Institute)

It will be noted that there is only a slight increase in power for pumping the two liquids. But, in the case of the viscous liquid, the capacity and head are lower and so is the specific gravity.

The example given in Fig. Add.15 can also be used to estimate the size of a pump as regards capacity and head on water when the duty on viscous liquid is known. The graph may be used for this purpose, since the results are an approximation in any case.

Assume a viscous liquid with the same characteristics as above, but duty point 250 m³/h and 43 m. The pump should be able to pump water at a capacity of 250/0.87 = 287 m³/h against a head of 43/0.94 = 46 m, assuming a pump efficiency on water of 85 per cent. A pump of these characteristics will be able to perform the required duty on the viscous liquid.

The power absorbed at duty point when pumping the viscous liquid is

$$P = \frac{9.807 \times 250 \times 0.9 \times 10^3 \times 43}{60 \times 60 \times 0.56} = 47 \quad \text{kW}$$

Figure Add.15 is one of a number of graphs designed to calculate pump characteristics when pumping viscous liquids. The Hydraulic Institute (USA) use an identical graph, as far as the lower part is concerned, but, for the correction factors, the efficiencies on water have not been taken into account. (The dotted lines are used for correction factors.) The present graph includes the peak efficiencies but, as mentioned above, they are not always applicable at the lower efficiency values.

It should be noted that if a two-stage or a multi-stage pump is considered, the head *per stage* should always be used in the calculations.

(h) Induced surges in pipelines

The analysis of surge conditions in a pipeline is complicated and time-consuming.

However, the reader should be aware that his choice of pump may affect surge conditions in the associated system, and from this point of view the salient factors will be considered.

The surges concerned here are often referred to as water hammer, and should not be confused with the term 'surge' as applied to variation in flow, such as occurs in the conditions outlined in (c) above when pumps with unstable characteristics are operating in unfavourable circumstances.

Surge is caused when the steady state of flow in a pipeline is disturbed for any reason. It is a transient condition lasting until a further steady velocity state (or state of rest) has been established. The degree of disturbance depends on the rate at which the velocity is increased or decreased, instantaneous changes causing high surge, while slow changes cause little disturbance of prevailing conditions.

If a sudden change occurs, such as a valve closing, the fluid behind the gate tends to build up in the restraining pipeline. The fluid is largely incompressible, and a shock wave is initiated which travels along the pipeline from the point at which the valve is located. This shock wave is reflected from the free surface of the fluid and travels back again to the valve. During the period of valve movement, normal pressures in the pipeline are modified by the algebraic sum of the transient shock waves emanating from the valve and those returning to it. Their severity depends on how quickly the valve is closing, the length, material, and thickness of the pipe, the compressibility of the fluid and rigidity of the pipe mountings. Upstream of the valve, the pressures are generally increased relative to the previous steady state pressures, and vice versa downstream.

In the same way that a valve closure changes velocity of flow to give rise to such surges, so also does any obstruction suddenly introduced into the pipeline. This is precisely the case when a pump, normally operating with manually-controlled discharge valves, loses its power supply. The pump immediately slows down and acts as a brake on the system. The effectiveness of the 'brake' depends on the inertia of the pump, motor, and other rotating parts. The resulting surge wave will depend on these inertia forces and the rate of change of the pump's characteristics as it slows down.

The tendency is always for a manufacturer to build to the lowest cost, and this must mean that the weight of moving parts is normally cut to a minimum. The rundown time of a pump on failure tends, therefore, to become small, and especially if a long pipeline with high velocities forms the system, then that system should be thoroughly checked.

On loss of power, the flow continues away from the pump and there is an immediate downsurge at the pump discharge which may, due to valve closure characteristics, become a pronounced upsurge later. Conversely, if the pump is subject to a long suction pipeline, the characteristic of surge on the suction side will be an immediate upsurge at the suction flange, with possibly a downsurge later as the valve closes. Eventually, the pressures on both sides return to the static pressures obtaining, a typical trace being shown in Fig. Add.17.

It should be noted, then, that the pump and valve should always be considered together, and that a suitable combination is one which reduces the maximum upsurge or downsurge to a minimum acceptable value.

In Fig. Add.16, if a pump fails, the fluid will eventually drain back through the pump, and in these cir-

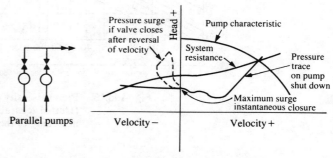

Fig. Add.16. Effects of flow reversal after shutdown

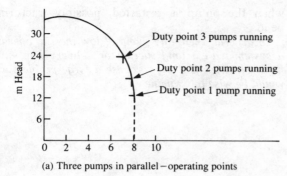

(a) Three pumps in parallel – operating points

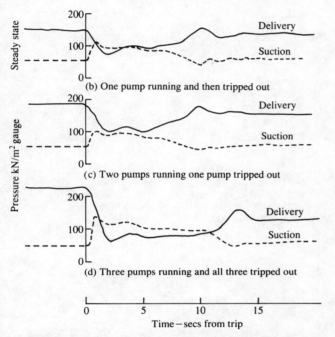

(b) One pump running and then tripped out

(c) Two pumps running one pump tripped out

(d) Three pumps running and all three tripped out

Time – secs from trip

Fig. Add.17. Surges on power failure of parallel running pumps

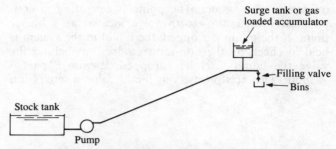

Fig. Add.18. Surge control at loading lines with quick-closing valves

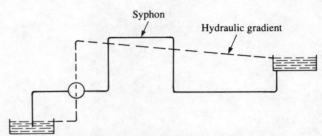

Fig. Add.19. Syphon on pumping line

near the filling process can reduce shock and noise considerably (Fig. Add.18).

Perhaps the worst effect of surge occurs when the downsurge caused reduces pressure in any part of the system to below the vapour pressure, and the liquid column separates as a result. Considerable damage can occur when the system recovers and the two liquid columns are rejoined. This can happen when the syphon principle is used in part of a pumping system (Fig. Add.19). The hydraulic gradient is below the top of the pipeline in normal working conditions (for example, where a pipeline is taken over other machinery or a road crossing perhaps) and when a downsurge occurs on pump failure the further loss of pressure may be sufficient to cause liquid column separation.

A similar effect is seen in Fig. Add.20 in the normal

cumstances the best time to close a valve is when this reversal is about to take place. Fast closing of a valve in such circumstances is not only acceptable but desirable, since leaving closure too late will meet increasing velocities in reverse. Surge in a system when the valve is closed at no flow would be confined solely to the effects of the pump running down, and this is a function of the inertia of the pump and its working head. The longer the run down time, and the greater the inertia, the less will be the resulting surge. It is, therefore, sometimes a cure to a surge problem to introduce a suitably-sized flywheel between the pump and motor.

Air vessels are often introduced into a system on the discharge side of a pump to minimize surge, advantage being taken of the compressibility of air to absorb the immediate shock effects by change of volume.

An example where precautions at the pump discharge have little effect is that of a 'loading line', where bins or cans are filled with instantaneous closure valves controlling the filling operation. In such cases, a surge tank

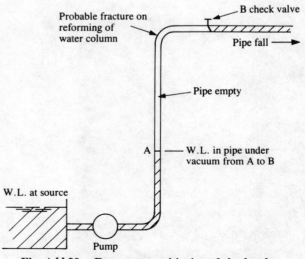

Fig. Add.20. Dangerous positioning of check valve

operation of a system. The pump is operating in conjunction with a non-return valve located at a higher point. If the pump is stopped, the liquid in the system is held in check by the non-return valve, but behind the valve the liquid level has dropped, leaving a partial vacuum. This acts precisely as liquid column separation when the pump is restarted, possibly fracturing the bend.

In general, then, any pump of low inertia, operating in a system with a long suction or delivery line, and with high velocities, is likely to cause a surge problem which it would be wise to examine.

APPENDICES

APPENDIX I. SYMBOLS AND PREFERRED UNITS

Function	Symbol	Unit	Remarks
Mass flow	W	kg/s te/h	Basic SI unit See note (2)
Volume flow	Q or q	l/s m³/h	Derived SI unit (m³/s $\times 10^{-3}$) gives convenient numerical values for the general capacity range of pumps (e.g., 25 l/s = 90 m³/h = 330 igpm). See note (2)
Density	ρ	kg/l	Derived SI unit (kg/m³ $\times 10^{-3}$) (a) gives convenient numerical values since most liquids have a density within the range 0.5–2.5 kg/l (b) numerically the same as specific gravity (c) useful in the direct equations: kg/s = kg/l $\times$ l/s cP = cSt $\times$ kg/l Operational unit te/m³ is identical to kg/l
Kinematic viscosity	v	cSt (Centistoke)	Derived SI unit (m²/s $\times 10^6$) kinematic viscosity is the funda- mental physical concept for most fluid systems, rather than the dynamic viscosity.
Temperature	T	°C (Celsius)	Basic SI unit taking zero as 273.15 K
Vapour pressure	P_v	bar	Derived SI unit (N/m² $\times 10^{-5}$) always given as absolute value
Gas pressure	P	bar	Derived SI unit (N/m² $\times 10^{-5}$) frequently given as gauge value
Head	H	m	See note (1)
Elevation	h	m	Basic SI unit
Power	E	kW	Derived SI unit, W $\times 10^{-3}$
Rotational speed	N	r/sec	Basic SI unit, preferred to the traditional r/min because it gives: (a) convenient numerical values mostly within the range 4 to 400 (b) direct relation with frequencies measured in Hz for vibration phenomena
Velocity	V or U	m/s	Basic SI unit

(1) For rotodynamic machines the generated head is defined as the energy imparted to unit mass flow of the fluid. The appropriate unit of head is then kJ/kg.

By regarding liquids as incompressible, we can substitute gH as the unit of head where H is the liquid free surface elevation in metres. The traditional convention is to neglect g as a constant and regard H as the pump head; this has the practical advantage that elevation and head can then be treated as equivalent.

For high-head pumps the compressibility of the liquid becomes significant and correction to the head H is required.

(2) Traditional units use time units of day, hour, or minute rather than the second. Their use introduces conversion factors into equations, inconsistent with the concept of coherence within the SI system, consequently the second is the preferred unit of time for calculations.

Hourly rates of flow are preferred for indicator display to operators.

APPENDIX II. RELIABILITY CLASSIFICATION

Installations having high availability are classified as follows.

Class 1

A Class 1 installation achieves high availability by having a single machine of high intrinsic reliability, and is characterised by:

(a) the machine being a single unspared unit upon which the process stream is wholly dependent;

(b) the plant section having a single process stream with a long process recovery time after a shutdown so that the loss of product owing to a machine stoppage is large even though the shutdown is for a short time;

(c) a capability of continuous operation within given process performance tolerances over a period of more than three years, without enforced halts for inspection or adjustment;

(d) component life expectancies exceeding 100 000 hours operation.

Class 2

As for Class 1, but where infrequent plant shutdowns of short duration are acceptable because the process recovery time is short. Consequently the period of continuous operation capability can be reduced and is taken as 4000 hours for this classification.

Class 3

As for Class 1 or 2 but with a machine performance deterioration during the operating period accepted, or countered by adjustment of process conditions or by other action on the part of the plant operators.

Class 4

A Class 4 installation achieves high availability by redundancy and is characterised by having:

(a) one or more machines operating with one or more standby machines at instant readiness at all times to take over automatically upon malfunction of a running machine;

(b) operating and standby machines designed specifically for their functions so that they are not necessarily identical;

(c) component life expectancies exceeding 25 000 hours operation.

Class 5

A Class 5 installation follows the Class 4 redundancy concept and is characterised by:

(a) one or more machines operating with one or more identical machines installed as spares to take over the process duty at the discretion of the plant operators;

(b) one or more machines operating in plant sections where product storage is sufficient to give the plant operators adequate time to assess the malfunction and take remedial action. Alternatively, plant sections where a single machine stoppage does not cause a disproportionately large process upset.

Class 6

Machines intended for batch or intermittent duty.

Where high demand availability is essential the machines lie within Class 4.

APPENDIX III. RHEOLOGY DEFINITIONS

1 Newtonian flow

Newtonian flow is characterised by the fundamental relationship

$$\tau = \mu \frac{dv}{dy}$$

where

τ = shear stress

μ = coefficient of dynamic viscosity

$\dfrac{dv}{dy}$ = velocity gradient (shear rate)

2 Time independent behaviour for non-Newtonian flow

2.1 Bingham plastics

Ideally, these have a defined value of yield stress. When this value has been exceeded the material behaves in a Newtonian fashion. Example: water suspension of rock particles.

2.2 Pseudoplastics

Their behaviour may be represented by a power law

$$\tau = K\left(\frac{dv}{dy}\right)^{n} \qquad \text{where } n > 1$$

For any value of dv/dy we may define an apparent viscosity, μ_a, such that

$$\mu_a = K\left(\frac{dv}{dy}\right)^{n-1}$$

Examples: polymer melt, dye paste.

2.3 Dilatants

Their apparent viscosity increases with increasing rate of shear and

$$\mu_a = K\left(\frac{dv}{dy}\right)^{n} \qquad \text{where } n < 1$$

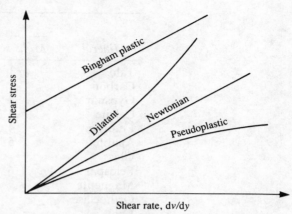

Note: The shapes characterise the fluid type, not the gradients.

Fig. App.III.1

Example: starch solution.

Summary. The relationship between these fluids is shown in Fig. App.III.1.

3 Time dependent behaviour of non-Newtonian fluids

3.1 Thixotropic

With these fluids, shear stress reduces with duration of shear at constant shear rate. The fluid structure progressively changes to lower the apparent viscosity. At rest, the structure can re-build to increase its apparent viscosity. Examples: inks, 'non-drip' paints.

3.2 Rheopectic

These fluids 'set-up' or 'thicken' rapidly when subjected to rhythmic shock or shaking. Example: gypsum slurries.

APPENDIX IV. HARDNESS VALUES FOR COMMON MATERIALS

Mineral	Hardness		
	Mohs scale	Knoop	HV
Talc	1	20	
Carbon		35	
Gypsum	2	40	36
Calcite	3	130	140
Fluorite	4	175	190
Apatite	5	335	540
Glass		455	500
Feldspar	6	550	600–750
Magnetite		575	
Orthoclase		620	
Flint		820	950
Quartz	7	840	900–1280
Topaz	8	1330	1430
Garnet		1360	
Emery		1400	
Corundum		2020	1800
Sapphire	9		
Diamond	10	7575	10000

Material	Hardness	
	Knoop	HV
Ferrite	235	70–200
Pearlite		250–320
Austenite, 12% Mn	305	170–230
Austenite, high Cr iron		300–600
Martensite	500–800	500–1010
Cementite	1025	850–1100
Chromium carbide	1735	1200–1600
Molybdenum carbide	1800	1500
Tungsten carbide	1800	2400
Silicon carbide	2585	2600
Vanadium carbide	2660	2800
Titanium carbide	2470	3200
Boron carbide	2800	3700

APPENDIX V. COMPRESSIBILITY OF LIQUIDS

The development of centrifugal pumps has narrowed the field of application of reciprocating pumps to duties where the pressures are high. Experience shows that the properties of liquids at such pressures are not readily available.

A helpful correlation of the volumetric properties of liquids was implicit in the work of Pitzer in 1955 **(1)** and others **(2)**.

An analytic equation for the compressed liquid region was derived by Tien-Tsung Chen and Gouo-Jen Su in 1975 **(3)**. This equation was simplified to

$$Z = (0.46711 - 0.83113 \cdot T_r + 0.54721 \cdot T_r^2)P_r$$

$$- (0.06685 - 0.19271 \cdot T_r + 0.14200 \cdot T_r^2)P_r^2$$

$$+ (0.00442 - 0.01307 \cdot T_r + 0.00969 \cdot T_r^2)P_r^3$$

where the compressibility Z is defined by

$$PV = ZRT$$

and P_r, T_r are the reduced (fractional) critical pressure and temperature respectively.

On a trial of 867 datapoints for 27 common petrochemical substances, the overall average deviation was 5.2 per cent and the maximum deviation 19 per cent (including the critical region).

References

(1) **K. S. Pitzer**, 'The volumetric and thermodynamic properties of fluids', *J. Am. Chem. Soc.*, 1955, **77**, 3427.

(2) **B. C. Y. Lu, C. Hsi, S. D. Chang,** and **A. Tsang**, 'Volumetric properties of normal liquids at low temperatures – an extension of Pitzer's generalised correlation', *AIChemE J.*, 1973, **19**, 748.

(3) **Tien-Tsung Chen** and **Gouo-Jen Su**, 'Generalised equations of state for compressed liquids – application of Pitzer's correlation', *AIChemE J.*, 1975, **21**, 397.

APPENDIX VI. INFLUENCE OF THE PROPERTIES OF THE LIQUID ON THE CALCULATION OF DIFFERENTIAL HEAD

The density of the liquid depends on both pressure and temperature. Now the temperature rise ΔT through each stage of a pump is given by

$$\Delta T = \frac{H_z}{102 \cdot \xi}\left(\frac{100}{\eta} - 1\right)$$

where

  H_z = differential head across the stage (m)
  ξ = specific heat of the liquid (kJ/kg · K)
  η = percentage hydraulic efficiency (per cent)

The density change due to this temperature rise, together with the change due to the compressibility of the liquid, is then used to calculate the differential head across the stage. Integrating these stage heads over the whole pump and referring the head to the inlet conditions gives the approximate total head.

For water, this procedure gives

$$H = 10.197(1 - 2.3 \times 10^{-5} \times P_d)\left(\frac{P_d - P_i}{\rho_i}\right)$$

where

  H = total head across pump (m)
  P_d = discharge pressure (bar a)
  P_i = inlet pressure (bar a)
  ρ_i = water density at inlet pressure and temperature (kg/l)

For pumps handling highly compressible fluids this procedure is inaccurate, and it is then better to calculate the head as the difference in enthalpies, but this procedure is beyond the scope of this publication.

APPENDIX VII. VISCOSITY CONVERSION TABLE

This table may be used for approximate conversion from one viscosity scale to another, at the same temperature. Note that all these measures are for kinematic viscosity, *not* dynamic viscosity.

Centi stokes	Saybolt Universal (seconds)	Redwood No. 1 (seconds)	Engler (degrees)
3	36	33	1.23
4	39	36	1.31
5	42	38	1.40
6	46	41	1.48
7	49	44	1.57
8	52	46	1.66
9	56	49	1.75
10	59	52	1.84
11	63	55	1.93
12	66	58	2.02
13	70	61	2.12
14	74	65	2.22
15	78	68	2.33
16	81	71	2.44
17	85	75	2.55
18	90	78	2.65
19	94	82	2.76
20	98	86	2.88
21	102	90	2.99
22	107	93	3.11
23	111	97	3.22
24	115	101	3.34
25	120	105	3.46
30	142	124	4.08
35	164	144	4.71
40	187	164	5.35
45	209	184	5.99
50	232	204	6.65
60	279	245	7.92
70	325	285	9.24
80	371	326	10.58
90	419	368	11.9
100	463	406	13.2
110	510	450	14.6
120	560	490	16.0
130	610	530	17.3
140	650	570	18.5
150	700	620	19.9
200	940	820	26.8
250	1160	1010	33
300	1410	1230	40
400	1870	1640	53
500	2320	2040	66
600	2800	2430	79
700	3250	2820	93
800	3700	3250	105
900	4200	3650	118
1000	4750	4100	133

Centi stokes	Saybolt Universal (seconds)	Redwood No. 1 (seconds)	Engler (degrees)
1500	7000	6100	199
2000	9200	8100	260
2500	11600	10100	325
3000	14000	12300	400
4000	18500	16100	530
5000	23000	20000	660

APPENDIX VIII. MATERIALS FOR SURFACE COATINGS

Material	Hardness (HV)
Electrodeposits	
Chromium	1000
Nickel	400
Cobalt Alloys	500
Thermally sprayed deposits	
13% Steel	330
18/8 Stainless Steel	170
Molybdenum	390
NiAl	300
Al Bronze	100–150
Stellite 6	300–700
WC/Cobalt	1300–1800
Cr_3C_2/Nickel Chromium	1100
Al_2O_3	2200
Cr_2O_3	2400
Spray fused deposits	
Nickel + BiSi	220–300
Nickel Chromium + BiSi	300–600
CrNi + BiSi + WC	450–700

APPENDIX IX. DATA SHEETS

(1) Data for use in this Design Guide are collected on:

    Pump Process Data Sheet 1 (Fig. App.IX.1)
    Pump Process Data Sketch (Fig. App.IX.2)

Pump Process Data Sheet 2 (Fig. App.IX.3)

(2) Data specified to pump vendors are entered on a Pump Data Sheet.

	PUMP PROCESS DATA 1		PUMP No. PROJECT No. Sheet of		
FLOWSHEET No. STREAM No.	PLANT	SECTION	WORKS AREA		
PUMP TITLE DUTY				System sketch separate sheet	

			Case 1 Normal flowsheet	Case 2	Case 3
1	Fluid common name				
2	Typical fluid composition	Wt %/Vol %			
3	Specific heat kJ/kg/C	Pour/Melting point			
4	Atmospheric boiling point °C	Flash point °C	TLV		
5	Highly flammable/toxic/corrosive/abrasive/odorous/unstable				
6	Materials recommended	excluded			
7	Running hours/year	Starts/year	Parallel pumps assumed Y/N		
8	Control by throttle/variable speed/stroke/bypass/press./elec/manual				
9					
10	Provide cases to cover operating envelope				
11	Temperature	°C			
12	Vapour pressure (bubble point)	bar a			
13	Viscosity	cP			
14	Density	kg/l			
15	Max/min temperature at standstill	°C	/	/	/
16	Solid content	% by weight			
18	Gas content	ml/m³ liquid			
19	Gas pressure in inlet vessel	bar g			
20	Level in inlet vessel	m			
21	Contents of inlet vessel	m³			
22	Friction loss in inlet system	m			
23	Gas pressure in delivery vessel	bar g			
24	Level in delivery vessel	m			
25	Friction loss in delivery system	m			
26	Loss across throttle control valve	m			
27	Differential head across pump	m			
28	Volume flowrate to delivery point	l/s			

ISSUE	PRELIMINARY	A	B	C	D	
COMPLETED BY						
AUTHORISED BY						
DATE						

Fig. App.IX.1

	PUMP PROCESS DATA SKETCH	PUMP No.
		PROJECT No.

PUMP TITLE

Give approximate line lengths, diameters and elevations.

ISSUE	PRELIMINARY	A	B	C	D	E
COMPLETED BY						
AUTHORISED BY						
DATE						

Fig. App.IX.2

	PUMP PROCESS DATA 2		PUMP No. PROJECT No. Sheet of		
LINE DIAGRAM No. CALC. No.	PLANT	SECTION	WORKS AREA		

| PUMP TITLE | | | | System sketch | |
| DUTY | | | | on Sheet 2 | |

			Case 1 Normal Flowsheet	Case 2	Case 3
1	Repeat of existing duty Project/Equipment No.				
2	Maximum flow limit	l/s			
3	Minimum continuous flow	l/s			
4	Momentary interruption allowed				
5	Significant running at no flow				
6	Suspended particles	wt %			
7	Suspended particle size range (micron)				
8	Suspended particle density	kg/l			
9			Jacketing: Tracing: Insulation		
10	Area classification Zone 0/Zone 1/Zone 2/Safe/Dusty				
11	Drivers Motor Steam/Hydraulic turbine				
12	Relief valve setting Upstream Downstream				
13	Piping flange rating Inlet Outlet				
14	Pipe lengths m/diameters mm Upstream		/ / /		
15	Pipe lengths m/diameters mm Downstream		/ / /		
16	Inlet strainer pressure drop m included in inlet friction loss Y/N				
17	Friction drop to bypass connection m				
18	Friction drop from bypass connection m				
19	Friction drop in bypass connection m				
20	Maximum permissible noise rating				NR/dB(A)
21	C/V stroke time < D/24 Y/N Narrow band or fast integral control Y/N				
22					
23	MACHINES DATA CONFIRMED				
24	Pumps in parallel Installed/Non installed spare				
25	C/V loss Case 1 m Bypass Flow				l/s
26	Shut off head m Pump type				
27	Noise Rating NR/dB(A) Casing maximum permissible pressure				bar g
28	Electricity volts			bar g	l/s
29	Water bar g			bar g	l/s
30	Steam bar g			bar g	l/s

ISSUE	PRELIMINARY	A	B	C	D	E
COMPLETED BY						
AUTHORISED BY						
DATE						

Fig. App.IX.3